Chasing Tales

by Ken Hopley

A different kind of travel book.

Grosvenor House
Publishing Limited

All rights reserved
Copyright © Ken Hopley, 2022

The right of Ken Hopley to be identified as the author of this
work has been asserted in accordance with Section 78
of the Copyright, Designs and Patents Act 1988

The book cover is copyright to Ken Hopley

This book is published by
Grosvenor House Publishing Ltd
Link House
140 The Broadway, Tolworth, Surrey, KT6 7HT.
www.grosvenorhousepublishing.co.uk

This book is sold subject to the conditions that it shall not, by way of
trade or otherwise, be lent, resold, hired out or otherwise circulated
without the author's or publisher's prior consent in any form of binding or
cover other than that in which it is published and
without a similar condition including this condition being imposed
on the subsequent purchaser.

A CIP record for this book
is available from the British Library

ISBN 978-1-80381-161-1
eBook ISBN 978-1-80381-162-8

This book is dedicated to my wife, Patricia, who has had the patience of Job for looking after me during my recovery since I had my strokes, as I would have killed me had I been her.

She has shown unflinching endurance, dedication and loyalty, as it would have been easier to look after a lion with a bad toothache that had not eaten a thing for a month, than a bad tempered bar-steward like me. THANK YOU, Pat.

Patricia and I in Montana, 1999.

Many thanks to my eldest daughter Elizabeth (Lil) for her devotion, skill and time in editing this book given her busy schedule, as I doubt very much that I could have done it myself.

Thank you Lil, XX

Thanks also to Ian Farrington for proofreading this book and to Kay Gallwey-Chand for the brilliant cover art work.

Contents

Introduction xi
Survival Courses 1
Egypt & Saudi Arabia, 1978 5
Bahrain, 1979 21
Basra Iraq 1st trip, 1980 31
Syria 1st trip, 1981 55
Ecuador, 1981 69
Korea 1st trip, 1981 81
Libya 1st trip, 1982 87
Baijy Iraq 2nd trip, 1983 99
Benin West Africa, 1985 115
Libya 2nd trip, 1986 133
Iran 1st Trip, 1987 151
Oil Rigs, 1988 169
Russia, 1989 183
Qatar, 1991 207
China, 1994 223
Gabon, 1994 235
South Africa, 1996 259

Nigeria, 1997 267
Azerbaijan, 1998 273
Egypt 2nd Trip, 1998 279
Norway, 1998 283
USA, 1999 287
Singapore 1st Trip, 2000 303
Iran 2nd trip, 2001 309
Korea 2nd Trip, 2003 319
Angola, 2004 329
Korea, 2006 335
Syria, Palmyra, Homs, 2009 347
Singapore final contract, 2011-12 353

Introduction

This is a book about my experience of 50 years working all over the world as an engineer, from major oil and gas companies (including the shady ones) to oil rigs, FPSO's refineries, gas plants and universities in some extremely interesting places, with highly interesting people. And yes, by interesting, I almost always mean odd. And sometimes just downright dangerous.

I was born in Walton, Liverpool, in 1947, which some might say was dangerous enough. After leaving school at 15, like most of us did, I trained as a marine engineer officer in Liverpool and started going away to sea. On leaving the Merchant Navy, I tried various and very mundane jobs in engineering in Liverpool but couldn't get ships and travel out of my head.

I tried various factory jobs in engineering that I found were too boring. Then I started working on pilot ships in the Mersey and Irish Sea. This was the beginning of a working life that wasn't what 'we all did', a life that took a young working-class lad to countries now vastly changed in terms of culture and politics - for better and for worse.

The result of all this is a different kind of travel book. The 30 chapters, each a different adventure in a different world, offer a unique perspective on a life. I retired earlier than expected, due to having had two major strokes at the age of 65 that failed to finish me off. By then, the job had tried to finish me off several times, from being bombed in Basra in the Iran/Iraq War of 1980-88 to being almost entirely awake during an appendicitis operation in Homs, Syria. From possible encounters with the KGB and the Russian and American Mafia to definite encounters with poisonous snakes - both the reptilian and human variety.

There are laughs amid the peril and of course a lot of booze. There is also some highly colourful language and my attempt to capture the syntax of the characters along the way, including my Scouse self, may not be to everyone's taste in our current climate. But this was a different climate, all 30 of them. I hope you enjoy them as much as I did. Well, as much as I did writing about them in any case.

Ken Hopley, 2022.

First seagoing trip to USA & Mexico, 1967. Posing in my new officer's uniform with purple Junior Engineer's stripes.

Same trip, different uniform. Bunch of engineering officers having a beer break during engine repairs (yours truly on far left). Temp on deck: 110 degrees F.

Survival Courses

A rebreather is a device which can be used in an emergency. You breathe out and expel the air, a third of which is C02, carbon dioxide, that is in your lungs, allowing you to rebreathe the expelled air. That's the theory.

So here am I again, strapped into a seat in a mock helicopter, waiting to be submerged in a swimming pool as part of a survival course. I had experienced this 'dunking' several times. Basically, you can't work on an oil rig without surviving a dunking. Mock helicopters, by the way, are basically a tin can with side openings representing doors and windows – not very realistic but you can escape quite easily. Just right for nervous people.

On this occasion, I was strapped next to a very nervous first-timer. As the 'old hand', I advised him to hold on to the window ledge next to him. Not a bad piece of advice as, when the chopper rotates upside down underwater, it's easy to get disoriented and forget which way is up. By holding the ledge, you can just follow your arm straight out of the open window. Again, in theory.

Of course a real helicopter has rotating blades, so after being submerged and rotated, strapped upside down in a seat, you are then told to count to seven, allowing time for the imagined rotating rotors to stop so as not to be cut in half. I doubt many, if any, students did that full count. I know I didn't on my first time.

On this occasion all seemed to be going according to plan. Splash, rotate, wait. 1... 2... 3... 4... 5... 6... 7. I'm just about to unbuckle my safety belt and get the hell out of the now-flooded chopper/bean can through my window to the surface, when what can only be described as a huge arse passed by my face. In a sheer panic, the first-timer had lost sight of his intended escape route and was going straight for mine! To stay alive, I now had to push this arse's arse through the window, a surmountable problem perhaps, had it not been for a couple more things... This particular mock helicopter in Singapore was actually less of a bean can. It had windows and doors fitted to make things more realistic (and more difficult) and they had added lashings of cold seawater to contend with too. Looking down, I could see the water quickly going over my ankles, then up over my knees, then chest high and coming up to my chin...

Time to activate the rebreather, I thought. I blew and exhaled into it as much as I could. Now completely underwater, I tried to breathe in. Nothing came out. It had malfunctioned. I was strapped to my seat, upside down, with cold salt water in my eyes, and had to find an arse-free door or window to escape through. Oh and did I mention it was dark as well? After what seemed like a lifetime, I unbuckled my belt with freezing cold hands. In moments like this, time stops. You are still upside down, totally disoriented and near blind as the cold salt water has hit you full in the face. The window is packed with arse, so you're now groping for a door. Then you have to find the locking mechanism (each helicopter has different types), open it, and in the meantime, try to take a breath, not remembering that the rebreather is malfunctioning.

Now the panic gets you. You are thinking: 'I am actually going to die,' but there's no time for a flashback through your life like drowning people get in the movies. The door miraculously opens and you go for the surface coughing and spluttering and try to look cool and make out that it was a piece of cake.

James Bond never had these problems. He ended up with a Martini, shaken not stirred, and a dolly bird at his side, as he gets his gun out and shoots the baddie before squeezing his latest squeeze.

Cough/splutter. Get dressed, then go for a pint, shaken and scared!

Welcome to the world of oil and gas.

Waiting my turn in the bean can (I'm in the yellow helmet in the back). ESSA Training Centre, Angola.

EGYPT & SAUDI ARABIA
1978

('Mr Kent' - unexpected surveyor, architect and camel victim.)

My contract was signed and my air tickets and itinerary had come by post. This was my first overseas contract and I didn't know what was awaiting me. I had been working as an engineer on the Liverpool pilot ships, a job that I loved and it was on my doorstep. The decline in the port of Liverpool soon brought the inevitable redundancies. Sad people with pretend happy faces due to the redundancy package that they knew wouldn't last were the norm over the remaining weeks. It was becoming ten green bottles as, one by one, personnel were laid off. I couldn't stand the lingering, the adieus and of course the drinking sessions as the latest victim was obliged to stand a round of drinks followed by a good few more and it felt like every Friday I was getting home half-pissed. I had to move fast as mortgage repayments have a habit of coming round too quickly for comfort. I started to apply for overseas work, even for jobs that I had never done nor had any experience of. It was: 'get the job first then learn it later.'

One of the positions that I had applied for, but didn't think I'd get, I got. It was at Jeddah International Airport as a maintenance planning engineer. So I had a great deal of trepidation due to a lack of knowledge of the job, the country and what to expect when I got there. I was on my way to Cairo, en route to Jeddah, the seaport on the Red Sea. I stepped off the plane and breathed in. It was like an oven but had the aroma of a Turkish men's toilet in downtown Istanbul (I would imagine). The taxi ride was a nightmare. I don't know what it is about some Arab taxi drivers but they drive as if being chased by a raging bull and wearing a diver's lead boot on their accelerator foot. Then they arrive at their destination, pick their noses and fall asleep. I was in the suicide seat and consternation was setting in. Cairo taxis do not have M.O.T.s! Basically, I was shitting myself. How do you slow this bastard down? "Swayer, swayer!" (Slow, slow!) came from the 'safety' of a rear seat. ("Swayer, swayer, you cupid stunt!" as the late Kenny Everett would say.) He grinned at me with a full set of bad teeth and dead-donkey breath.

At the hotel, our spirits lifted as the bar was still open and we could laugh about our situation. We were a team of seven, our agent called us his Magnificent Seven, but he was blinded by his commission. One of the guys was a rotund painting inspector from Blackpool, who had worked in the Middle East before and seemed to know the score, which was handy. I took an instant liking to him; he was full of jokes, hearty and good-humoured and nothing seemed to bother him. I later found him to be very good company, which was needed on this project as, away from home, you can soon find yourself on a downer. This guy had the ability to make light of a bad situation and would soon have you laughing when you met him. The world could do with more people like him. As the hotel was full, I had to share a room with a supercilious old fart. Every time somebody laughed he would tut, tut, and would say, "Bloody children." I was lucky that it was only for one night.

The flight to Jeddah wasn't until late afternoon, so I took in the Cairo museum and was chuffed to see the Tutankhamun exhibition. In the mummys' room, you could actually see their teeth and hair. I found myself alone looking at some of the wonders of the world, when a hand shook my shoulder, which I am sure didn't do wonders for my heart. It was one of the Seven.

"We're off to see the Pyramids."

"That sounds good, wait for me."

The Great Pyramid, I must admit, was quite awe-inspiring, but I found the Sphinx a disappointment as it was a lot smaller than I had envisaged and had been cordoned off and surrounded by scaffolding due to restoration work being carried out. Its face was being eaten away by dreaded pollution which was getting a hold.

"Mister, mister, you want ride my camel?"

"No, go away."

"I no charge." This sounded like a good deal. Having never ridden a camel round the pyramids before, or in fact anywhere else, I was captured. The ugly, smelly beast crouched down, followed by the camel. I was tempted to step on his back and up onto the beast but he was only picking up something that he had dropped. The camel protested, totally annoyed that I had dared climb on board. I was whip-lashed back and forward quite violently (it would make good astronaut training) then off we went and, after about twenty yards, the camel driver said: "You pay now?"

"You told me you no charge."

"Yes, that is correct. I no charge, but camel, he charges. He has to eat, pay the rent, TV licence." I thought, 'You naughty boy,' but couldn't help but smile at the cheek of it. What else was I expecting? Nothing is free in this world, especially a camel ride round the pyramids. After some bartering, I managed to get away with giving him about two pounds sterling. But I did get a hat off him and took some good photos. I had read about a young lad in Barnsley who was suffering from leukaemia, who collected hats of every type, so I sent him the camel driver's hat and a photo of me wearing it next to the pyramid. He sent me a St Christopher medal.

I packed, showered and set off for the airport. Cairo airport was then just a bit better than a Nissan hut and hot as hell. I had a most welcome cold beer or two, our last ones before I returned to Blighty, or so I thought. The plane was a Boeing 747, a Saudi Airlines Jumbo, full of men dressed in dishdashas on their way to Mecca to become hajjis. Every male Muslim tries to make this once-in-a-lifetime pilgrimage. I was bursting for a pee and went to the toilet and opened the door without the engaged sign displayed to find a man standing on the seat having a number two. I found all the toilets were all full of these chaps standing on the seats number two-ing. I returned to my seat and, as I walked down the aisle, had to step over some pilgrims who were sat on the floor. To my horror they were just about to light a Primus stove to make tea. I thought, 'Oh shit,' as you do when somebody is making tea, right over the 747's wings, which contain the fuel. I asked the French stewardess, who was almost out of her mind with the tribesmen, to help me stop them. Most of them had probably never seen a car before let alone a huge aeroplane. To our relief, somebody could speak Arabic and between us, we managed to avoid what could have been a disaster. I could see a typical Sun headline: 'Jumbo jet blows up over brew up in mid air.' Going through customs was a sheer delight. I watched a Saudi customs officer searching a young Indian who was wearing the ubiquitous flared trousers, wide-collar pastel-coloured shirt and Gary Glitter platform shoes. He wanted to search his shoes, so what did he do? He hammered the poor guy's (probably only) pair of shoes on the side of his desk until the heels came off, then just tossed the now-destroyed shoes in a bin and told him to go.

We were driven to the company guest house and settled in for the night, thanks to the lack of any bars. The next day we ventured into town, found a cafe and sat outside drinking fresh orange juice, when the wailing started.

"What's that?" we said in unison. All the shops were closing, shutters were being hastily shut and people were running each and every which way, as if an air raid had started. It was prayer time and it was the Moi'zan calling all Muslims to prayer. What should we do?

"Just sit here until it goes away," seemed a good suggestion. We were sitting minding our own business, when a Muttawa (religious policeman) started shouting and bawling at us – for what, we didn't have a clue with this being our first time in Saudi Arabia. He went away but soon returned with a real policeman, who immediately knocked all our drinks off the table with his swagger stick and pulled out a gun.

"Oh shit. I think he is telling us in his polite way to fuck off!" We had been sitting drinking orange juice, minding our own business, which had infuriated the Muttawa because it was their prayer time. We didn't know what he expected us to do. It is incredible what freedom we (and many foreigners) have in the UK and yet you can't look sideways in some of these countries. The things we do for oil and contracts.

Back at the company guest house I was on the balcony, looking down and recognised a face from the past. It was Thelma, my sister's childhood best friend from Norway! What a coincidence and what a small world we live in. We used to play together when we were kids. She recognised me and we both said each other's name in unison. I couldn't believe it. I shot down to have a chat and a huge catch-up. It must have been at least thirty-five years since we last saw each other and she had hardly changed. She had married an American pilot, had two children and was based here, whilst her husband trained Saudi pilots in Jeddah. Brave man.

"What are you doing tomorrow evening?" she asked. "We're having a dinner party, I'd love you to come and meet everyone and talk about our childhood." Invitation accepted!

Next day we had to go into the office for all the necessary registrations, visas and handing over of our passports. I didn't like the idea of my passport being taken but what can one do? The office was full of pompous men, all wearing long white dishdashas, looking like ghosts, bossing people about like school children. That evening, I went round to Thelma's earlier as I couldn't wait to get out of the boring guesthouse.

"Sorry I didn't bring any wine. Off-licence was closed."

A joke as Saudi Arabia is a dry country.

"No problem, we have loads of booze here," said Thelma's husband, asking me my preference. Phew. He told me that he was a pilot with the American air force and was now on a two-year secondment pilot-training program with the Royal Saudi Air Force. The other guests started to arrive. They were either pilots, or from the American embassy, hence the reason for the abundance of booze. During dinner, there was an explosion in one of the cupboards, followed by another, then another. "Shit!" Thelma's husband exclaimed, as some of the home-brew beer had started to pop their tops. It turned out to be a very interesting evening, lots of good conversation and laughs. I offered to babysit for them, which turned out to be a good arrangement, as I could escape from the guesthouse and watch a movie on TV, have some beers and good food.

The next day, I found to my dismay that my colleagues had departed for other parts of the Kingdom and I was stuck in Jeddah on my own. My first day at work, I was introduced to my boss, an American. He greeted me and told me that I was the new building surveyor.

"No that's not me. I'm here as a planning engineer."

"You sure you're not a building surveyor?"

"Quite sure."

"Shi-it.' Which I am sure is one of America's favourite expletives. And becoming one of mine too. We went to see the base manager. I sat down in his office and the ubiquitous chai arrived on the scene. Tea is always served in small glasses and

very sweet. You are never asked if you want sugar or not, it's always assumed that you like half sugar / half tea, mixed.
"What is the problem?"
"The problem is, I am not a building surveyor."
"But you are engineer, you can do this, no problem."
So that was that, I was now the new building surveyor at the Royal Saudi Air Force Base.

I had a team of craftsmen/tradesmen working for me, of all nationalities under the sun. One was a giant carpenter from the Yemen and there were plumbers, builders, electricians from India, Egypt, Africa, locals and every Tom, Dick and Harry. They couldn't pronounce my name Ken so used to call me Mr Kent after the cigarettes, which made them laugh. I just got on with the job in hand, surveying buildings as best I could, because I had decided that it was pointless to argue. So I went round surveying the buildings; the roofs, walls, gutters, cracked toilets (yuk). I used to abhor inspecting the toilets, as I could never hold my breath long enough. One had to feel sorry for the plumbers – what a shit job! I would then go back to my office and write out the work orders for the work to be carried out, easy-peasy.

My office window happened to be facing the main runway and I could see all the aeroplanes landing and taking off. What's that goal doing there? Surely they don't play football by the runway? I was soon to find out it was where the local butcher slaughtered... whatever he slaughtered. One day a goat had caught the smell of the butcher's apron and bolted, with the butcher in hot pursuit. I suppose that is why it is called a runway. I could just imagine a British Airways pilot coming in to land and seeing that scene up ahead: "I say, what the eff is that?" One of the main runways was under repair, when an Egypt Air jet came in on the same runway. It landed, bounced over excavated soil, chopped some poor bloke's head clean off and careered down the runway, the plane that is.

All credit to Egypt Air how quickly they evacuated the plane with minimum carnage. Pity about the poor guy who had an unexpected hair cut that day.

One day, I walked into the one of the billets, just as lunch was being served. I very foolishly said "Kwais" ("good" in Arabic) so they beckoned me to join them in their goat dip meal, full of everything - and I mean everything. I tried desperately to get out of it by patting my tummy trying to make out that I was full but they were not having it. I put my hand in the bowl and tried to bypass the eyeballs and made a stab for the rice, hoping that the rice hadn't been too near them. My American boss, who was married to a Thai woman, used to go to the Thai canteen for his lunch every day. One day he took me. I thought, 'I'll just have a bowl of chicken soup, can't go wrong with that.' Until I saw the innards of the chicken, oh and a foot thrown in for good measure - and must not forget the skin. I just drank the soup. The Thai chef came over to us, spotted that I had almost finished and gave me the Thai equivalent of a thumbs-up and, to my horror, topped up my bowl again. The Thais, like the Chinese, will eat anything. Never stand still by them or you're in the pot.

Another thing that used to amaze me in this country were the abandoned cars by the side of the road. Some were almost new, a Mercedes that probably just had a simple problem wrong with it, even a flat tyre. The Saudis would simply abandon them and get another, rather than taking the trouble to fix it.

Jeddah was getting me down with me being on my own, plus the job was so boring so I requested a transfer, telling my boss that, "I hadn't come to Saudi Arabia to work as a building surveyor." It's all very well inspecting cracks and gutters but to do another professional job that you are not trained to do is a definite no. As nobody could do my job who hadn't had the training either, I eventually got my transfer and headed for Riyadh, the Saudi capital. I boarded the Saudi Airways flight and found myself sitting next to a Saudi woman dressed in their national costume. She had one of those beaked masks on her face and stared at me for most of the flight as if I was from Mars. I pity Saudi women. I don't know what it is like now but in the seventies in Saudi Arabia, you would never see women on the streets, or in shops, and they never worked. It was a shock to see Arab women go into the sea fully clothed, head scarves on as well. The secretaries were nearly always Indians, Thais or Filipino men. I assume the women must have been prisoners in their own homes.

I landed in Riyadh, went straight to the head office and was greeted by another American. "Hi, Ken, been expecting you, you're our new building surveyor?"

"No!"

"Shi-it, but that's what it says here."

"I am a mechanical engineer, here to set up a planned maintenance programme, I am not a building surveyor."

"You god-damn Limeys are a pain in the ass."

I was losing my temper now. "Look, I can't help it if the company has made a mistake and hired the wrong person!"

Our tempers now subsided and we both calmed down. He looked at something on his desk. "Why hey, we got a position in Kay-mus-moo-shi-ite."

"Where?" He meant Khamis Mushait, which was North Saudi, up in the mountain region. "Yeh, maintenance engineer supervisor."

"Sounds OK." I would have agreed with any job as long as it wasn't a bloody building surveyor.

I flew to Abba, where the base was and drove the remainder of the way to Khamis Mushait. I entered the main office and this time it was a Saudi who greeted me. A tray of chai appeared, with the usual, nossa noss (fifty-fifty) sugar tea mix.

The dentist must do a roaring trade in these parts. He prattled on. "I am in charge of this base and I want it to be the best in the company." I nodded in agreement. "I would like you first to give me an artist's impression and then an estimate of the cost."

"Pardon? Excuse me, but what position am I assigned to here in Khamis Mushait?"

And no word of a lie, he said: "Architect. Quantity surveyor."

As I picked myself up off the floor, he was still blathering on about swimming pools and leisure centres. I left his office with an overwhelming feeling of utter doom and was shunted into another office to get indoctrinated with the company's policies, etc. The superintendent was a huge black American, who, after shaking hands and sitting down, immediately fell asleep in the middle of the conversation. I sat there, wondering what I had said that was so boring. He woke up with a start, carried on indoctrinating, then fell asleep again. 'What is wrong with this guy?!' He woke again and apologised, then put my mind at ease for a brief moment by informing me that he suffered from sleeping sickness. 'Great,' I thought, as he was going to drive me around the place and this was a mountainous region. By some miracle, I managed to survive the ordeal as you may have guessed and he didn't fall asleep at the wheel. Allah must have been watching over us.

I had now made my mind up that come tomorrow, I was going to resign. I got to my room and thought of all the time that I had wasted on this job. I took a quick shower and went for dinner. Who should be in the dining room, but the painting inspector from Blackpool. I was never so glad to see someone in my life. I told him of my situation; he laughed and said, "No problem, just do it." The next day in the office, he called me over. "Hey, Scouse, look at this." He had found a letter in his desk from a guy from Louisiana who was asking his friend to get him fixed up with a job. It went something like this:

'Howdy, Herb, can you get me a job in Saudi Arabia? An engineer's job? I've had lots of experience back home fixin' and mendin' things for ma around the farm. Be nice to meet up agin! Ma and Willy send their greetings to you. You take good care of yourself now, Jake.'

After we had stopped choking, I sat at my desk and thought, 'Sod it'. I was not too bad at art in school and fixin' and mendin' things. So I set out to do an artist's impression of the swimming pool and recreation facility. I must admit, I quite enjoyed doing it. Artwork completed, I popped in to show the Saudi base manager. Well, he was over the moon with my drawings. "Kwais, shukran. (Good, thank you.) Please start with the swimming pool and give me a price for it, we should be able to start the civil work soon." Artist's impressions no problem, but how do I get a price for it? Luck was with me as there just happened to be a Swedish base next to ours that had a swimming pool. I went to see their camp boss and explained my predicament. He showed me around the pump house and filtration system. I made copious notes and sketches of the plumbing arrangements and he gave me the names and addresses of various suppliers. "Fuck them, stick any price down, they have more money than sense," was the advice from the painting inspector.

I felt that I had to be a little bit more precise than that so I spent the week visiting suppliers, getting prices and delivery dates for pumps, filters, plumbing, cement, tiles and even tables and chairs for the pool. I somehow managed to get a price together and the Saudi manager didn't bat an eyelid. "Told you didn't I?" the painting inspector said. Next day, the manager wanted to see me. "The King is coming here next month and I want to impress him with our base." I nodded, dreading what was coming. He continued, "I was thinking of an archway over the road in front of the base." 'Oh shit,' I thought. It was getting into the realms of fantasy here.

"An archway? A permanent structure?" "No, no, I mean, something like an advertising hoarding." I could see it now, the archway falling on top of the King as he passed under it and me being beheaded or jailed for life. The archway was the next project while I waited for the civil work to start for the swimming-pool complex. For the artwork, I incorporated the company's colours, alongside the Saudi coat of arms. The manager was delighted and immediately ordered it to be made. I never saw it built or erected I am very glad to say. The King survived and so did I, so I reckon it must have been OK.

I was about to prod a chip, when I heard this awful wailing coming from the kitchen. "Oh god, no, no, not John, Johneeeey!" Steve, the company secretary, had just heard on the radio, that John 'The Dook' Wayne, had just ridden off into the sunset for the very last time and he was hysterical. "Silly c*nt," the painting inspector exclaimed. It was interesting times. You could be sitting in the TV lounge and a slimy male local would walk in smelling of old Old Spice. You would then hear a door click open and the bottle of aftershave would disappear into Steve's quarters. You could hear all the sound effects through the wall. As our base housed the American F15 jet fighters, the security was supposed to be tight but it was nothing to see a security guard throw his rifle to the ground to hug and kiss a friend. I was driving through the main gate one day and was stopped and made to get out of the vehicle. I had the correct pass with my photograph in it. My passenger had a borrowed pass but it was me that got arrested and questioned.

This place had its odd characters, one of which was American Bill. I didn't quite catch what the hell his job was but once a week he had to drive out into the desert to pay a visit to some of the local farmers and inspect their water wells to make sure that they were clean and had no dead goats or whatever in them, as we were getting our supply from them. One day

I drove out with him. Now the tricky part was, we would have to tell the farmers (who were also tribesmen and were brandishing swords and knifes) that we would stop their floose/money if they didn't 'get their act together.' Easier said than done.

"I say, old chap, would you kindly remove that dead goat out of there, old bean, otherwise we will stop your money."

"OK, mate, no problem, will do." Said as he withdrew his sword from his belt.

We went inside the farmer/tribesmen's house, which was round and surprisingly cool. It didn't have any stairs, just a long slope going to the top. The ubiquitous tea came out, "50, 50 please." I noticed a wad of riyals not-so-surreptitiously being handed over to Bill. Of course. Hillbilly Bill was the moonshine maker. I should have guessed, with him coming from Mississippi. He was supplying the locals, a dangerous game to get into out here, especially supplying, as they would drop you in it as soon as the heat was on and, unknowingly, I was now an accomplice to Billy-boy's dastardly deeds!

One of the Americans brought his wife out on a short trip and the Saudis put on a party for her. It was just like a kindergarten party, with balloons hung all over the place - party hats, cakes and lemonade. We did a deal with the Thais for some Sidiki (moonshine) as we didn't want to get involved with Silly Hilly-Billy and his un-surreptitious dealings. We took the drink into the party in the guise of Coca Cola. As the night wore on, we were beginning to enjoy ourselves (getting pissed) when the local police chief arrived unannounced, wearing all his kit, with swords criss-crossed on his chest and a gun at his side. Definitely not a man from the Met. 'Oh fuck'. And what does the crazy pissed jock do? He starts dancing and getting the bloody police chief up! How we survived, I will never know.

The time was going quite quickly, what with being busy and I was due for leave. I got to the airport and was usurped by a

Saudi, who got my seat. This meant that I would miss my connecting flight to London and I would need to change my ticket. Oh, how I wanted to kick him up the arse but not recommended in his country. I ended up having to return to the base and couldn't believe the sight that hit me. It was just like an absurd front cover on one of Tom Sharpe's sidesplitting books, with water spouting all over the place. The police, using a JCB, had dug up a water pipe and had got stuck down the hole that they were digging. Thais were being frogmarched and kicked up the rear end and the Saudi police had sprung into action. What had happened was this: Thai contractors had been gambling and getting well-pissed on their lethal home-brew, hence a knife produced by a Thai was inserted into the stomach of a fellow Thai, who I assume must have been cheating. He then sprung a leak. The Saudi police had been called in on the leaking Thai incident and the Thais were now helping the police with their enquiries. The Saudi plod were hell-bent on finding the demon drink, never mind the mad knife person, or the dying Thai with a hole in his belly. 'Shit.' I shot to my room and emptied all the beer from my fridge down the loo and got rid of all the bottles. After doing all that, they didn't search my bloody room.

Ex-pats who have spent time in Saudi Arabia soon discover that there is a generous quantity and a variety of alcohol available, even though it is absolutely forbidden to manufacture, possess, sell, carry, or drink. It could be of excellent quality, nearly hangover-proof and could be manufactured in almost any household kitchen. The preparation of such alcohol, (Sid) Sidiki, or 'Juice' as some people called it, was a revered secret, yet eagerly always shared with good friends. One large oil company had produced an alcohol-making manual called 'The Blue Flame' on how to make just about any kind of drink. It also contained drawings and instructions on how to manufacture a still. As certain trail-blazers had almost blown themselves up in the process of manufacturing them and also in the process of processing the said 'Sid.' Historically, alcohol distillation began with, strangely enough, the Arabs. I bet many of them don't know that! However, they did not invent, but greatly improved the cumbersome methods used by the Greeks in distilling turpentine from rosin. The Arabs were the first to distil wine and our wood alcohol is plainly Arabic in origin. The present prohibition in Saudi Arabia is sternly enforced, especially by adherents of the Islam (Muslim) religion.

"Were I to begin the world again, with the experience I now have of it, I would lead a life of real, not imaginary, pleasures. I would enjoy the pleasures of the table, and of wine, but stop short of the pains inseparably annexed, to an excess of either."

> The Blue Flame – A Treatise on producing
> "Hooch" in Saudi Arabia.
> American oil giant, Aramco.

The Thai contractors had craftily buried plastic bins full of Sidiki under the ground outside their accommodation, then put sheets of plywood on top and covered them with soil, where they grew flowers. Good eh? It had completely fooled the police. The Thais had been the cause of my further delays.

Around this time, a colleague arrived from Jeddah and had some bad news for me that bowled me over. Thelma, my childhood friend that I had recently met in Jeddah after all those years, had passed away. She was only in her mid thirties, she was not overweight, and in fact she was on the skinny side. I immediately thought of her two small kids and her pilot husband. Suddenly my cancelled leave paled into insignificance. I left Saudi Arabia with somewhat mixed feelings. I had intended to return but gave it a wide berth. I would never go back there for all the Sidiki in China. Or in fact, anything.

Me being taken for a ride. Literally.

Giza pyramid & Sphinx, Cairo.

BAHRAIN
1979

(A Sniper, some Jock fights and a pickled egg.)

I arrived at Manama airport minus my luggage. The airline had mislaid it - in other words, some arse at Heathrow hadn't loaded it onto the plane. Those of you who've been through this experience know what it's like and those that haven't can imagine. It's one hell of an inconvenience - no change of clothes, no toiletries, reading material etc! By the time I arrived at the company guest house, I was not in a very good mood. At least I had my duty free... There was a Jock from Burntisland already installed there; he'd been in Bahrain for two days. We chatted away for hours and got to know a bit about each other. I was to be his boss. After several glasses of vodka, I soon forgot about my problem, at least until the next day.

The project was a gas plant and we had the role of inspectors, overseeing the installation, pre-commissioning and commissioning of Mitsubishi gas turbines. Although we got on great with the Japanese engineers. They didn't like us checking their work and hated losing face, even over the smallest things. We found little notes such as: 'Sorry for having to borrow a torque wrench from the warehouse' and 'Could we please witness the torque wrench settings.' Arriving on site minus such a critical tool as a torque wrench was unheard of from a Japanese employee. Their uniform was very... uniform to say the least - all dressed the same, just like the military, with the company logo all over the place. I am sure they were stamped on their bottoms as well. The staff of the company Banagas (known in Bahrain as 'Banana Gas'), along with the main oil company, housed their personnel in a small township called Awali. We used to go for our lunch at the 'greasy spoon', the Banana gas restaurant. They had their own bank, supermarket and cinema - the latter I visited once and once was enough! The place was a blend of farts, sweat and cheap aftershave and the floor was covered in sunflower seed shells, as the locals would sit with their feet up on the seats in front, eating the seeds and spitting out the shells all over the place.

Jock and I were joined later by another ex-pat from Lancashire, who reminded me of a Japanese sniper, as he wore those bottom-of-a-milk-bottle glasses. It was always welcome to have another person in the team to spread the load and conversation but this guy turned out to be a right little shit! For a start, he didn't like the idea of me, or anyone, being his boss. One day he was inspecting what is known in the oil and gas industry as a fin-fan, a fan that cools the gasses that pass through a huge radiator, similar to a car radiator. The only difference is, this particular fan was twelve foot from tip to tip and had six blades. I'd given Sniper the job to check the blade clearances. Being his first day on site, I popped over to see how he was getting on. I asked him if it was isolated. "Not necessary," he replied, even though it was requirement on the permit. I walked past the start button which was sticking out like Schnozzle Durante's nose (look it up), available for anybody just passing to accidentally bump into. I knocked the button with my shoulder to demonstrate this and 'whoosh', this huge fan was in motion. They can be started from the control room by an unwary operator and are mainly set to start automatically as the temperature rises. I should have made sure myself that the system was isolated but he was there and I had given him the job. A lesson learned but I gave him a bollocking anyway. In future, he was to make sure the equipment was isolated and locked out as he would not have a future. I had no wish to retrieve his head and have the unenviable task of informing his wife that the once head of the family was now headless. Also, and a very big also, I was supervisor and did not wish to end up in jail.

The Sniper used to mark his bottle of whisky so that Jock and I couldn't help ourselves to it, not that we would... well, speaking for myself that is! Most weekends, Jock and I used to go into Manama town but he'd just sit in by himself, which to be fair, was his own prerogative. Thursday night was the equivalent of a Saturday night in the West. We called a taxi instead of driving into town ourselves, as the local nick was not on my wanted list. We arrived at a night club in Manama, paid the driver and told him to pick us up at two am. The club was frequented by many airlines crews. After a while we spotted a pissed Geordie operator from the plant, zigzagging his way over to some Gulf Air trolley dollies and was inviting one of them to have a dance. I could see her head doing an Indian yes, which is the opposite of a Western yes. I read his lips: "Well, I suppose a blowjob is out of the question then, love?" Oh shit. Fortunately for him, she saw the funny side, her being a Brit, but I don't think the burly American pilots did!

We left the club on time and sure enough the taxi was outside. We arrived back at the house, paid and tipped the driver, only to find that he wanted more money, considerably more! He seemed rather pissed off. We gave him some more, and he still wasn't satisfied. Jock told him to go away in Scottish and was soon helping himself to the Sniper's whisky, when we heard a knock on the door. "Who the hell is that?" It was three o'clock in the morning. I looked out of the window: it was the police. Jock's tribal instincts soon took over; he came to the door brandishing a large kitchen knife.

"Christ, Jock, you're going to get us locked up for life. Let me answer the door!" He hid behind the door as I asked the policeman what he wanted. "You come to police station," he said, beckoning to his car. "What for?" He didn't answer. "No, I come Bukra (tomorrow)," I replied and shut the door. He knocked again several times but we didn't open it. We thought no more of this until the next day coming from work. As we turned the corner of the street where our house was, we saw a police car parked outside, obviously waiting for us to show up. Oh shit. We drove past looking the other way. I noticed the Sniper was grinning, knowing full well that we were in a spot of bother. "OK," I said, "let's get it over with." They took us to the local nick where the chief of police was waiting for us. He made Jock and me go into separate rooms to write out our statements from the previous evening. This is what had apparently happened... We had told the taxi driver to pick us up at 2am and he had waited outside the night club all the time we were in there, which was about two hours! The police chief read our statements and realised the situation but we still had to cough up to the driver. The chief said, "You can afford it, he cannot." How can one argue with that? Being only too glad to get out of there, we paid the driver, shook hands and left. That's one Bahraini police chief's logic and another lesson in the school of life!

Sheikh's Beach. Now there's a place. The Emir of the island is a small fat guy, with a turn in his eye, probably caused by ogling the women from his beach house. There was a rumour going round at the time that if he took a certain fancy to a woman, he'd send one of his servants over to invite her to take tea with him on his veranda. Then the next step was an invitation to a party. At the party, he'd hand out Rolex watches or whatever to his grateful guests but – and I say this was only a rumour but it was a well-known rumour - he didn't give Rolex watches out for nothing. Some of the stories you'd hear from the aviators would make your hair curl, or fall out. Air stewardesses would tell us sorry stories about how some of them were virtual prisoners in their apartments and were treated like slaves 'etc' by their Bahraini paymasters.

One day on Sheikh's Beach, I got talking to one of the crew of another well-known international airline. I won't say which but he was the co-pilot of a 747 and was flying out that very night, having been drinking in the afternoon - how scary is that?! I should have informed the airline. Jock and I had also arranged to go to Sheikh's Beach the following day with some of his mates. They were late. We waited but they didn't show up, so we decided to call round to their apartment. The front door was open so we walked in. I couldn't believe what confronted us. There were bodies lying all over the place, on the floor, upside-down, bent over armchairs... The Jocks had obviously had a bender last night! Empty bottles of beer, whisky, and vodka were scattered all over the place. The place stank like a Glasgow pub on a Saturday night. Quite astonishingly, they seemed to sober up in an instant - amazing Jocks. However, they didn't want to go to the beach but to the rugby club where we ended up drinking those vast cans of Foster's lager, the ones that you needed two hands to hold. I felt a bit peckish as I foolishly hadn't had breakfast before I started drinking, and got myself a pickled egg. I took one bite and it got stuck in my throat. I began to choke, and with all the Jocks being smashed they didn't notice me dying. I bent over the back of a chair to do the Heimlich manoeuvre to myself and out popped the egg. "You OK, Jimmy?" was all I got.

After a rapid recovery, we shot off to a bar in Manama where they had some pool tables. We ordered drinks and waited our turn, after the many locals and Filipinos playing and gambling. I noticed one of the Jocks saying something to one of the Bahrainis, in a not-too-polite way (a sort of late Saturday night in a Glasgow pub way!). Apparently, what had happened was, Jock's head was in the way of a Bahraini who was going to play a shot and he'd been told to 'move' in an arrogant manner. Well, Jock turned round to the Bahraini and said, in a typical Scottish manner, "Speak to me like that again, I'll knock your effing heed off, Jimmy." I spotted the Bahraini running to the phone (this was before mobile phones came on the scene). As I was the only one reasonably sober, I went over and asked what the problem was. As if I didn't know. He told me that he was an Arab prince (aren't they all?!) and he was calling his friend the chief of police (oh no, not him again) to get us jailed. We all swiftly piled out of the place and got into our vehicles.

"Where's Tony?"

"He must be still inside."

By this time, all the Jocks were getting tooled up with bottles, iron bars, all the Saturday night Sauchiehall Street necessities - basically anything nasty that they could lay their hands on.

"Shit, look, give me five minutes," I said to them. "I'll go back and get him out."

"Five minutes and we're comin'."

This could easily turn very nasty, I was thinking. If someone got badly hurt or killed, what would the consequences be? Execution, or life in a Bahraini jail? I got to the door and could hear shouting and bottles and glasses being smashed. I opened the door. Tony was slumped over the bar completely oblivious to everything. With that, a bottle flew over my head. I could see blood had been splattered on the once-white walls. Luckily, he was by the door, so I grabbed him and dragged him outside. We got to the car and shoved him inside. Just before we drove off, four police cars came screeching into the car park with their sirens on and lights flashing. We waited for them to go inside and shot off. Lucky or what?

Drink can obviously be a problem for ex-pats and I am no angel. You are away for sometimes long periods from your respective families, homesickness gets you as life can get very lonesome and very stressful. When you start a new contract, especially commissioning which comes at the end of a project that's been running for maybe a year or two, you are expected to get up-to-speed ASAP and hit the ground on a motorbike. So you may worry about the job and or how long will it last. You also could have the worry of the country. Take for instance, Nigeria. Now that is a very scary place to get about without being robbed or murdered!

The weekend was on us again and Jock and I called in on a colleague from our plant. He lived in Awali, the company town. I rang the bell several times but no answer. The door was slightly ajar so we went inside. I could smell something burning and went into the kitchen only to find eggshells all over the floor, walls and ceiling. The pan was red and on a gas ring that was full on. In the hallway, I noticed blood on the walls. 'Shit, what's happened here?!' The chappy in question was fast asleep at the kitchen table, pissed out of his mind. I woke him. One side of his face was scratched as if a big cat had attacked him. I cleaned him up and made sure that he was awake before we left him. The following day at the office, he accused Jock and me of duffing him up until I pointed out to him the blood on the wall. The wall had a stippled, sharp Artex effect and he had fallen against it, and had slid down using his face as a brake. That man had problems.

We came across locals with pots of money, living in an expensive villa and the outside in the street would be a rubbish dump. I assumed that they didn't care about their immediate surroundings. Imagine if, outside your front door, there was a tip or leftover rubbish from the builders who had built the house. One would be slightly cheesed off, but here they just don't care so long as their villa was OK! A contractor would install an air-conditioning unit and yet not complete the surrounds. A hole would be knocked in the wall but you'd end up with a space around the unit and could see outside. Sort of defeats the object somehow of having air conditioning! I once went into a local police station to apply for a driving licence. They had installed a telephone extension to an adjacent office and instead of just drilling a small hole to accommodate the cable, they had knocked a hole in the wall big enough for Mike Tyson's head to go through. You would also find that an outside wall would be built snakelike, to follow the contours of the land. This meant that at some stage, where the wall ended, it was at an angle of 45 degrees and for sure, would eventually fall down.

Jock had decided to resign and was leaving at the end of the week. He asked me for a loan of £300 as he wanted to buy some presents; he didn't have a credit card and could only give me a cheque. You have to trust somebody sometimes. It didn't bounce - I didn't expect it to! To see him off, myself and the Jock clan headed into town. Of course the Sniper stayed at home with his whisky. We arrived at a place called Mansouri Mansions in Manama. Each of us ordered a club sandwich, the ones which should have a danger sign on them because of those bloody toothpicks that nearly always end up going through the roof of your mouth and out through your nose. Soon the drinks started to flow Jock style. As the night wore on, my mate Jock had gone to the toilet. He'd been gone a while and was not missed until someone asked where he was, as nobody had seen him for ages. Problem was he had to leave now for the airport to catch his flight and the silly sod was missing. We searched everywhere inside the place, even the women's toilets, but he was nowhere to be found. I noticed the time: it was 23:00 hours and getting near the 'Ashura festival' time, where the local Muslims flagellate themselves with knifes and chains, then fall into a zombie-like trance and walk about covered in blood, children as well, with their parents' blessing. Just like at Old Trafford, when Manchester United play, 'er' clash, with Liverpool. The religious ceremony comes round once a year, which I suppose is rather fortunate for some! However, it would not be the ideal time for any Westerners to be out and about. There was only one place left - the car park! We dashed outside and found him lying face down. I turned him over and examined him. He was OK but very pissed and in a mess, covered in mud and dust. Miraculously, all his money, wallet, passport, gold chains and kilt were still on him - obviously nobody had seen him. I drove him to the airport, cleaned him up as best I could, then got him through check-in. We said our goodbyes and I left to drive through a crowd of zombies walking past the car, staring without seeing, covered from head to toe in blood. I don't even think they saw the car, let alone me. I didn't see Jock, or hear from him again until we bumped into each other on a project in Africa years later.

I was now left with the Sniper. A shiver went down my spine at the thought of it, as Jock and I had become friends, as much as you do on a foreign contract, and used to spend many an hour upon the flat roof of our house, having a wee dram and putting the world to rights. A team of operators from UK, who had only been on the plant a couple of weeks, had a dispute with their agent and decided to do a runner, a midnight flit. They had left without any notice or warning! Of course the Sniper was quick onto the scene as they'd left everything in their apartment, food that is, and Snipes wanted to take it all back to our place, which I suppose wasn't a bad idea, as we were both bringing out our families out for a holiday. We cleaned our apartment until it was spick and span in anticipation but as time went on, the job just got much worse and I found that I just could not live in the same house as this guy, with his pathos mentality. With that, I too resigned. It annoyed me when I think of all the supplies that I helped gather and all the cleaning that I did. 'Still, that's life!'

Me and the Sniper in the desert (I'm on the right, radio station in background).

BASRA, IRAQ, 1ST TRIP 1980

(Life and death in the Garden of Eden.)

"The Arabian Gulf is the arsehole of the world and Basra is fifty miles up it." This was really comforting and reassuring as I was about to go there! I was in the Pig and Whistle, the British Embassy bar in Damascus, Syria, in conversation with one of the staff. We were discussing the Middle East situation and Basra in particular, Iraq's second largest city and my next port of call. The conversation got onto flying, when an American chipped in.

"I hate goddamn flying!"

"Have you done much?" I asked.

"It's just that I've had so many motherfucking bad experiences."

"Like what?"

"Got myself shot down in Nam and had to eject. The Vietcong nearly shot my ass off while I was trying to escape."

"Oh. I see your point."

I had just just recently started taking flying lessons. My instructor was an off-duty police inspector from the Liverpool vice squad who, in his spare time, was an unpaid flying instructor and did this in order to keep up his flying hours (you need seventy-two hours per year to keep your licence up-to-date). We took off and headed towards Liverpool city centre and a bit too close for comfort to the Merseyside police headquarters. "Just giving the lads a wave." As he flicked the wings up and down, Biggles-style (the fictitious pilot hero, written by Captain W.E. Johns, whose books I used to read as a boy). We then headed to the training zone, away from the built-up areas. I had warned my wife and kids that I may be flying around our house at approximately midday. We were heading in that direction and I spotted our road and said to the instructor, "Any chance of flying down there? That's where I live and I can see my house." I pointed to it. He, in a definite command, said he was taking over the flying, as you do when swapping over at the controls, and we were soon circling the house sideways.

When I eventually opened my eyes, I could see my wife and kids in the back garden, jumping up and down and waving; they seemed about half-an-inch tall. After doing two, maybe six circuits (I didn't count, as it's difficult with one's eyes shut!), the instructor did his Biggles bit, waved the wings and then we shot out to sea, following the coastline and back to the airport again. When I arrived home, I was full of it of course, telling everybody about my experience, when our next-door neighbour rang the bell. The first thing he said was. "Did you see those idiots flying over our houses this afternoon?" He'd been out washing his car. My wife pointed to me: "That was him!" "Oh sure, pull the other one." He didn't believe us and thought we were joking.

Before I got too involved with this flying school, since it took up most of the flying time getting to and from the training zone, I decided to change to one that was nearer my home and the training area and not too far from the Royal Birkdale golf course. At the new school, I was stopped and checked in by a security guard whose face seemed familiar. It turned out to be the ex-Everton goalkeeper Gordon West, who was in their 1966 FA Cup-winning team. Bloody hell, what a comedown. When you think of the difference in today's game and the nonsensical and highly inflated salaries. After the classroom briefing, we headed towards the plane which had to be refuelled. It was a Cessna 150. We then checked the fuselage, landing gear, ailerons, propeller etc, then climbed in and checked instruments, joystick, movement of rudder and ailerons, and shouted "All clear!"

I started the engine, then taxied towards the runway, parked up, did a final instrument check and waited for the tower to give us final clearance. I sat there waiting with bated breath. "You're clear for take-off," came the instruction from the tower controller.

"Go ahead," said the instructor.

"Er, do you know that this is only my second flying lesson," said I.

"Do you want to fucking learn to fly or not?" said he.

"Er, yes."

He grabbed my hand and placed it on the throttle, then we began to accelerate.

"See those trees in front of us?"

"Yes."

"If you take your hand off the throttle we will hit them and die."

I could feel the plane lifting. Unbeknown to me, he had already primed the craft for take-off. "OK, pull back slowly, now." I eased the stick back and we were airborne. What an incredible feeling. Every two to three hundred feet into the climb, I levelled and dipped the wings from side to side, to enable us to see if there was any other aircraft in our flight path. This was a training area for the RAF and Liverpool university cadets, a rather crowded and dangerous training area for rookies, I thought. The flying school, which is also an RAF base, had on display by the entrance a Phantom jet, which used to send shivers down my spine (you'll find out the reason for that later on in this chapter).

Sorry about that, went off on a tangent a wee bit. Now where was I? Basra. I arrived at Saddam International Airport, Baghdad, around midnight and ended up having to spend the night in an upright chair. The next day, I flew down to Basra on an Iraqi Air internal flight. Apart from the odd one or two Westerners, the remainder of the passengers were Iraqi. Feeling rather groggy after a sleepless night, I flopped into my seat and tried to get some shut-eye, when I spotted a hen. What the hell did I have to drink last night? I rubbed my eyes and had to pinch myself. It was still there, on the lap of an Iraqi sitting next to me. He had brought a bloody hen on the plane and was sitting there with it on his knee as if it was the most natural thing in the world. 'What's going to happen when the food is served?' I thought. As the flight wasn't too long they didn't serve any. I couldn't sleep anyway because I was worried in case this fecking hen started pecking at my pecker.

We flew over the Marshes, the home of the Marsh Arabs, which looked truly amazing. A driver was waiting for me at Basra airport to take me to the camp. I hate living on camps but this one was quite good. I shared a bungalow with a fellow Scouser who, to my delight, was also an Everton football supporter and could argue all night with some Liverpool supporters, also working at the plant. I don't know how he managed it when you consider what Liverpool FC had achieved over the years! The plant was a huge petrochemical complex and I was the new maintenance supervisor. My boss, Mike, showed me around. Half of the plant had already been commissioned and the other half was in the pre-commissioning phase, so it was a mixed bag and so were the crew: Brits, Americans, Arabs, Poles, Muslims, Christians, drunks, skivers, the lot. The storekeeper was from Bangalore, India, and reminded me of Spike Milligan so of course his nickname had to be Spike. I was 'The Red Baron' because I used to wear red Red Wing American coveralls on site.

The camp was huge. It had two swimming pools, an off-licence, two bars, launderettes, hairdressers, two restaurants, a bowling alley and two supermarkets. It even had a nightclub, which had been built out of old packing cases. When you were inside after a few drinks, the ambience was so good that you'd seriously think that you were in a nightclub back home. The walls were painted black and they even had one of those revolving mirrorballs on the ceiling. The bar had all the spirit optics and was well-stocked. We each had a monthly drinks allowance for the off-licence and, if you wanted to join the club, you'd have to hand it over to the bar committee to keep the drinks flowing. On my first weekend, I went on a pub crawl to the two bars and ended up at the nightclub. I later tried to find my way home but, with all the bungalows being the same, got lost. I heard some music coming from one of them and knocked to get some directions. An American couple, whose bungalow it was, were having a party. They took pity on me and invited me in to join in. We became good friends but some time later, things ended tragically for this lovely family. I also had my share of tragedy that year. It started with the sudden death of my brother-in-law, who was only in his early forties.

The job was going very well, it was interesting and I was busy all the time. We had our share of problems that had to be solved and rectified, shutdowns to plan and execute and, on top of all that, we still had to commission new equipment. Our operations supervisor, my opposite, was an American/German who used to strut into the control room like Captain Kirk in Star Trek and would survey the control panels. It was amusing to watch when he did this; it was such John Wayne bullshit but it would look good and convincing to any newcomers to a plant like this one. I had previous experience, so he didn't gain my admiration or awe!

There were various characters on site, as always. One, an Egyptian, used to pester me for safety boots as the contractor he worked for wouldn't provide any. He continuously reminded me that he was a Christian and would show me his crucifix. What that was supposed to make me do, I don't know. I gave him an old pair of my site boots, Red Wing type, which are quite expensive. Of course he painted them yellow and red, which made him look like a clown, which he was. The ungrateful bugger was indicating that he'd expected a new pair. Of course, if you give one person new boots, then all the other contractors would want them and you wouldn't get any peace. Another character was a sly Iraqi, who was always skiving off to have a smoke, usually in a non-smoking area. I never actually caught him at it but I'd constantly see him without his hard hat on and would give him a warning. Twice I had to bring him into the office with an interpreter and explain to him the dangers of smoking and not wearing his hard hat on site and the consequences of what would happen if I caught him again. I'd get into trouble, as he was one of my men. Then I caught him again. Enough was enough and I was going to fire him. He pleaded with me I gave him one more chance and he understood that if I caught him one more time he was history. I hated having to fire people, especially poor people, because I couldn't help but think of their children but then again, if the breadwinner got killed, it'd be worse for them and of course the other poor sods that he would take with him, like me.

There were always safety issues to contend with. Some of the operators were from Canada, ex-farmhands with hardly any training or experience. One day, I had technicians working inside a vessel, when an operator opened a valve, even though the valve was safety-tagged and locked closed. He started to fill the vessel with water and almost drowned them. We not only had a language barrier but also an idiot barrier. On the plant there were several modes of transport: three-wheel bikes were a hit as you couldn't drive a vehicle in most parts of the plant due to making sparks in a volatile area, etc. One day,

I saw this young Iraqi labourer trying to ride one; he kept going round in circles. "What the eff is that silly arse doing?" He only had one eye, hence him going round in circles. Not long after his cycling circling, he was called up into the army to fight. Sad thing is he's probably dead now. Can you imagine the gift that he would have made for an Iranian sniper. "Shit, missed him. Never mind, here he comes again..." Bang.

One weekend, we planned a trip to the marshes which I was really looking forward to. We took two huge cool boxes full of beer and food and set off at dawn. After the usual hassle and haggle with the local fishermen over money, we hired a boat and crew and set off. It was a beautiful day. I sat at the stern of the boat. It was so silent and peaceful, water buffalo were chewing into their breakfasts and an occasional boat would pass by, going the opposite direction, loaded with reeds. The woman would be doing the donkey work rowing the boat and, as per usual the lard-arse male would be sitting at the pointed end, sleeping!

This was sheer bliss and I had my first beer of the day. We eventually arrived at a lagoon which was a truly amazing sight. The houses were made of reeds and had been built on their own floating reed islands. This was the home of the Marsh Arabs. The children came running over to greet us. When I stepped off the boat, I began to bob up and down, which seemed weird as it's usually the other way round! We met the chief, leader, or elder and were invited to his house for tea and to take some bread. I had borrowed my boss's Polaroid camera for the trip and started taking photos. The villagers would stand upright as if on parade, then smile, until the moment you pressed the button, so it came out with them looking dead serious. They obviously had never seen an Instamatic camera and were puzzled at not seeing anything on the photo. When it started to develop in front of their eyes, the look on their faces was a treat. Of course everybody wanted a photo. I gave some out – one to the chief, naturally. The children were exited and each child wanted one, but I only had twenty. There was one little girl about six years old, standing at the back who pleaded with her eyes. I bypassed all the grabbing hands and gave her a photo. Her face said it all. I still have her photo to this day, of her silently waving and watching us depart.

This geographical area is where the Tigress joins the Euphrates, the Delta. As our guide showed us around, I couldn't help but notice how incredibly peaceful it was; you could hear a pin drop on the Moon. The guide stopped at a tree. 'What's this about?' I thought.
"We are now standing in the Garden of Eden and this is the very apple tree!"
"Are you kidding me?" I found myself saying, as I am neither deistic nor theistic! I cannot know for certain but I think God is very improbable and I live my life on the assumption that he is not there. But I maybe sightly deistic as I think somebody or something must have started things before the Big Bang. Right or wrong, who knows?! I don't believe in old men with long beards sitting on clouds, or fairies at the bottom of my garden, or that the Earth was made in seven days and someone switched the Sun on and there was light! But we must have come from somewhere and someone or thing must have pushed the button.

So here I was, standing in the Garden of Eden as described in the Book of Genesis, where the first man, Adam, and his wife, Eve, lived after they were created by God, the garden that forms part of the creation myth and theodicy of the Abrahamic regions. The creation story in Genesis relates the geographical location of both Eden and the garden to four rivers: Pishon, Gihon, Tigris, Euphrates and I was standing on the Tigris and Euphrates delta. I personally do not believe in the story but you could still have knocked me over with a feather.

I remember the song 'The Garden of Eden' by Frankie Vaughan, from when I was a lad in Liverpool. Whether one believes in all this or not, it was still a remarkable place.

We walked on and eventually came to a building. Inside were photos of this strange craft, made of, yes you guessed, reeds. It was the Tigris, Thor Heyerdahl's third boat. He had constructed it here in 1977 in this very place for his Tigris expedition where he'd set off on a journey to find out whether the world's oldest civilisations have a common source. Heyerdahl believed that "Somewhere along the way modern man has gone wrong by failing to acknowledge the huge difference between progress and civilisation." He built a boat from reeds grown in the marshes of Iraq and embarked at Basra in Southern Iraq, uncertain as to the route he would take. He sailed down the Persian Gulf, now Arabian Gulf, to Muscat, then across the Arabian Sea to Karachi and back across the Indian Ocean to Djibouti, where he and his crew were refused permission to go any further for 'security reasons'.

Heyerdahl decided to give the Tigris a symbolic end and sent a message to the UN which said:
"Today we are burning our proud ship with full sails, and the rigging and hull in perfect condition, as a protest against the inhuman elements of this world of ours. Our planet is larger than the bundles of reeds that bore us across the sea, but still small enough to risk the same threats, unless those living acknowledge that there is a desperate need for intelligent co-operation if we are to save ourselves and our common civilisation from what we are turning into a sinking ship." I myself have always thought: 'what if we destroy this planet of ours, which no doubt we are, if we carry on in our selfish ways?' It is 2021, we have Coronavirus. The Earth is like a ship, the difference being we do not have lifeboats (another Titanic!)

Despite its dramatic end, Heyerdahl's expedition proved the possibility that the civilisations in Egypt, Mesopotamia and the Indus Valley, now modern Pakistan, which sprang up almost simultaneously and which shared all man's major inventions, could have been linked and could have inspired each other. This was mind-blowing stuff. I can understand the travel writer Gavin Young, the author of many wonderful books and one in particular, 'Return to the Marshes', about the magical experiences he'd had living here amongst these lovely people. I often wonder what ever happened to them after that evil bastard Saddam Hussein had drained the area, one of the wonders of the world probably gone forever.

After one of the most memorable and enjoyable days of my life, I came down with a bang. On my return to the camp, I heard people were looking for me. I went to my boss's house where he broke the news to me that my mother had passed away. From bliss to grief in one day. This was the moment since my childhood that I had always dreaded. No matter what age you are, when your mum dies, the shock and grief is awful.

The next day I was on my way home. I had to stop at the Kuwait border and, as I reached the immigration office to get my exit visa, one of the guards shut the window in my face and I was the only one there. He had shut shop so that he could get some sleep. To think what we did for them in the Gulf War, saving them from 'Mad Sad' or… who am I trying to kid? I think the word 'oil' came into the equation somewhere; why else would the West want to save them? I was stuck there for over two hours waiting to get my passport stamped. By the time I arrived at the airport, I had missed the flight and had to go through the hassle of changing my air tickets and waiting hours for the next flight to London, with not a bar in sight.

The funeral took place the following day. The usual crowd were there: aunties, uncles and cousins that you never see except at a funeral. Richard Matthewman mentioned in his charming and very witty stories about his childhood in Barnsley about people always having an uncle Albert, as did Del Boy and Rodney in 'Only Fools and Horses'. In my case it is very true, as I do/did have an Uncle Albert, a real one unfortunately. After the funeral, we went back to my brother's house for the dreaded family reunion. I indeed have one of those families that only meet up at funerals. "Well, see you next time, then; who's next on the list?" Distant relatives that I hadn't seen since my childhood were there, including my Auntie May who used to suck crisps, as she thought it terribly working class to crunch them and make a noise. I remember her from when I was a boy. She would turn up with Derek, a namby-pamby cousin of mine, who had a halo above his head. She would ask him repeatedly in front of everyone. "Have you done your Jobbies?" Jobbies was her posh name for poo. She was married to a clerk and lived in a house no bigger than a postage stamp but they had a garden, which made all the difference. It was about the size of a grow bag.

Auntie May and my other aunt, Dolly, used to look down their noses at my mum because she had married a mere 'lorry driver,' who was below her station! My uncle, the one married to Auntie Dolly, was also a driver but he happened to work for the Royal Mail and drove a post van, now there's class for you. First class! Pun intended. It must have been the Royal crest on the side of the van that did it. Strange thing, snobbery, isn't it? Uncle Tommy and Auntie Joan were also there. Now we are talking big-time posh. He had reached the dizzy height of 'clerk of works'. She worked behind the counter in a cake shop, which was no ordinary cake shop, as it was a 'by appointment to Her Maj' cake shop, Crest and all! Classy Cake co. No scallywags, tramps, mere lorry drivers or common working classes allowed. They hadn't had any children, they just collected those boring Lladro figurines as substitute children. He used to call round to our house when there was a match on at Goodison Park and park his motorbike in our back yard. My big treat of the week was to sit on it and pretend to be riding in the Isle of Man TT races.

And then there was good old Uncle Albert – nearly forgot about him, pity that! After his demobilisation from the army after the war, he must have thought, 'Right, done me bit, now to sponge.' And so he copulated and populated and fathered god knows how many children, most I don't think I ever met. He had never done a day's work in his life as he had reversed the process of retirement, retiring before he worked! I suppose he never had the time, with producing all those sprogs. He never had any money either, well, only for booze and cigarettes. Everybody, and I mean everybody, used to look down their nose at Uncle Albert and he didn't give a toss. My dad coined the phrase 'lazy Scouse git' long before TV's Alf Garnett in 'Till Death Us Do Part'. Auntie Elsie, his long-suffering wife, worshipped the ground that he walked on, no, the chair that he sat on, and laughed with a phlegmy rasp at every word he said. At my mother's funeral, I happened to be standing by the food table, when he staggered drunkenly past, reaching for the cheeseboard with his hand in a certain 'je ne sais quoi' way, then grabbed a huge chunk of Stilton, stuffed it into his mouth and immediately spat it out. "That tastes like shit!" he exclaimed, then he put the remainder back on the board and staggered off to get more booze and to annoy the clerk of works and the Royal van driver. I made a mental note not to go near the cheese.

Albert's sons, my cousins, were still in the Teddy-boy era stage and this was 1981. It was as if they'd been locked in a time warp. They sported long sideburns that went down to their necks, tight trousers, long drape coats, suede shoes and probably had a razor in their pockets. The undertakers called at the house for their customary drink and tip. One turned out to be an old school mate of mine. He told me that he was only standing in for a mate and had a much better job. (It seems the general pattern that if and when you ever bump into any old school mates, they have all got super jobs.) He was in a hurry to go and see Liverpool play Manchester United at Old Trafford, and I thought, 'You bloody turncoat,' as he used to be an Everton supporter until Liverpool started to win a few cups and things. My two sisters were there: the eldest, Jean, with husband number..? I honestly can't remember, as she'd had more husbands than Liz Taylor and Zsa Zsa Gabor combined! My other sister, Betty, pale as usual, was with her roly-poly husband, who had worn surgical gloves for years - dermatitis was the problem, poor fella. My two brothers were there: Alan, whose house we were in, was trying to accommodate everybody, mopping up after Uncle Albert, making tea for Auntie May to dunk her crisps so as to avoid making any working-class noises - with her little finger sticking up from the teacup at the ceiling, like a submarine's periscope, she drank her tea looking down on the riff-raff. My elder brother, John, who was known to everybody as Sonny, was not looking too well. He had recently been diagnosed with Hodgkin's disease. Alan and I used to take him for a drink at least once a week. One night we were in a pub, Alan had gone to the toilet and there was just Sonny and me. He looked me straight in the eye and said: "Am I a goner, kid?" Unbeknown to him, I had been to see his GP who happened to live across the road from where I live. He had told me the full extent of what my brother had. I felt so helpless. I looked him in the eye and said, "Don't be bloody daft, oh, it's your round." He wasn't stupid, he knew. The following week, he asked me if I could get him some stepping-stones from the beach, for his garden. I drove to the beach, searched for and loaded the car with these stepping stones and I knew that he would never use them.

I flew back to Baghdad. My leave was just three weeks away and had made arrangements to meet my family in Cyprus.

The day before I left, I received a phone call from my wife, telling me that Sonny had passed away. I was so used to him going into hospital for chemotherapy and always bouncing back, I never dreamed that when I returned to Iraq, I would never see him again. My wife was ready to cancel our holiday. I said, "No, we go ahead as planned." She had been looking after our son, Richard, who is severely disabled and needed the holiday. Both of us were stressed out to the limit, plus my son and daughter needed it too. Knowing my brother, he would have told me to "go ahead" in his true buccaneering style. I went ahead as planned and arrived a day before my wife and kids and had to check into another hotel because ours wasn't booked until the following day. I felt extremely morose and alone, so I got pissed. I know that is not the answer to your problems but I couldn't think of a better one at the time.

My family arrived and we went to check into the hotel. Our travel agent had told my wife that, as she had booked the hotel and I was flying in from Baghdad, I would be able to bunk in at no extra cost. When my wife told me this over the phone, I thought that it was a bit pie-in-the-sky and had my doubts but went along with it. They couldn't and wouldn't fit me into a room with my family. When is this hassle going to end? I could have murdered the travel agent. We ended up having to book a suite, because that was the only vacant room that the hotel had available. I found it very difficult to enjoy myself for obvious reasons but I had to make some sort of effort for my family.

My wife had spent the afternoon paragliding and came back beaming; she had recently made a parachute jump for charity and didn't mind the heights. Even though I like flying, I can't stand heights. What the hell, I thought. So I had a go. I was attached to the boat by a rope and strapped into a harness. The boat roared off, giving me just enough time to run a few yards and up I went. The boat suddenly looked about the size of a fly and I was looking down at the tallest of the beachfront hotels. I started to sing, trying to stay calm and was just starting to enjoy myself, when the boat started to turn round to go back. I started to come down, thinking, 'What is the moron doing?' I was only about forty foot above the sea and he was going flat-out. Oh I see, he's playing silly buggers and going to skim me off the water. I hit the water at about thirty miles an hour, which is pretty quick when you are suddenly dunked in the sea being dragged behind a bounding speedboat. I was just like a pebble when you throw it sideways in the sea and it bounces. He was still going flat-out and had not noticed my predicament. He eventually stopped, but I'd been bouncing and twisting about so much, the rope I was attached to had wrapped itself around me and I started to sink. I remember looking up as I was spinning and sinking. The sky above me was getting darker...

I had to do something quick, like a scene from a James Bond movie. I unbuckled the harness but the rope had wrapped around my legs and had me trapped. As the chute was full of water it was dragging me down. I don't know how but through shear desperation I managed to scramble to the surface and shout to the boat driver but, to my horror, I started to sink again. I thought, 'It's curtains time.' I struggled and fought like hell and managed to get to the surface again and he was there. I grabbed the boat. I think if I'd sank again that would have been me 'finished with engines,' as we used to say on board ship. Because the weight of the chute was too much and I was battered, tired and felt like I had swallowed half of the Mediterranean. I clambered on board coughing and spluttering. Unbeknown to me, my wife and some of the hotel guests had witnessed the near debacle.

That night we got ready and went down to dinner. Over pre-dinner drinks, I had my leg pulled over what had happened.

Someone said, "Would you do it again?" "Certainly, no problem," said I, after a few drinks and the old Dutch courage had taken over my senses. The next day, as we negotiated our way around all the German towels to get to the pool, there was a small crowd of our newly found pre-dinner drinking partners, armed with cameras. "What's going on here?" "Don't you remember what you said and promised last night?" my wife said. I felt a sickening feeling in my stomach. "You told them you'd do it again." Oh fuck. "There he is!" they cried out in unison. There was no way out, I was committed: I had to go up again. The walk down to the beach was, I imagined, probably the nearest thing to being led to the gallows as one could possibly get. The mob followed, baying for my blood. The boat owner started to tighten my harness, rather a bit on the tight side, I thought. "Why," grunt, "did you," grunt, "take off" grunt, "the harness?" he asked, grunting. "Because I had to dive for it in deep water and ended up with a nose bleed," I was tempted to say. "What did you want me to bloody do? Drown, you idiot?" but thought, 'Better not,' as my life was, yet again, in his hands. So I said, "Sorry," as we British do.

We started off and I was soon airborne and felt OK, piece of cake. The fear had subsided. I started waving to people, who seemed like ants below me on the beach. I found myself whistling in that smug way and thinking of all the free drinks that I'd now won by going up again. We approached the turning point and a sudden feeling of déjà vu and apprehension took over my smugness. I was going down again, same spot, and the boat was going flat-out same as before.

'Oh shit, not again.' About forty foot from the sea, I started to unbuckle the harness and was going to ditch it, and drop into the sea. No sooner had I opened two of the buckles when I started to go up again. 'Oh fuck.' I desperately tried to buckle myself in again – luckily there were four straps, and two were now undone. For his revenge, I'm sure that the bastard took me over all the yachts that were in the harbour, as I looked down at all the sharp masts pointing towards my arse! I finally plopped gently in the sea, ordeal over. "Get the bloody drinks in, you lot!"

We had had a wonderful time on the island and it was over too soon. My wife couldn't find the air tickets and we ended up having to go to the airport without them. A big problem at the airport and just what was needed after a vacation. Air Malta and immigration eventually let my family board. As my flight back to Baghdad wasn't until the following day, I headed back to the hotel and had a quiet day by the pool to reflect on things. After a leave, I always feel totally depressed and severely homesick and wonder what the hell I am in this business for. After a few days it wears off but it's a feeling most people working abroad in similar situations will tell you.

The job was interesting and I had now moved to a place of my own by the swimming pool. Mail was getting through fine as there was always somebody going on leave and somebody returning, who would take or bring letters. My wife and I used to send each other tapes. I used to love getting back to my room after work and sit by the window having a beer and listen to the tape. One day I recorded a tape whilst lying on my bed having a drink or two but fell asleep and, lucky for me, I played it back and immediately had to tape over it, as I'd unintentionally got pissed and recorded complete rubbish.

I had to go into Basra to get my visa updated. On the drive into Basra along the main road, I looked out of the rear window of our truck to see what was making such a dreadful noise and found myself looking down the barrel of a gun that happened to be attached to a tank that was right behind us. And behind the tank was a convoy of army vehicles going at a fair speed. The people in the immigration office seemed to be behaving strangely, huddled together in quiet conversation.

I got my visa stamped and headed back to the plant. The remainder of the morning, I saw fighter planes flying around at a high altitude. Trains were going by carrying tanks and other armaments. This was the start of three days that I will never forget.

It was lunchtime and we drove to the restaurant, when all of a sudden, two jets screamed over our truck at rooftop height. A "what the fuck was that" comment was said in harmony. We were witnessing a dogfight in the sky above us. An Iraqi MiG was chasing an Iranian Phantom. Every conversation was rife on what we had seen. After lunch, we headed back to the site and on the way we counted at least twenty Iraqi MiGs that were taking off from a nearby airfield, fully laden with bombs and missiles. 'What's going on?' kind of went through my mind! All afternoon we saw more Iraqi MiGs flying overhead, more trainloads of tanks and armaments going by. After work, everybody headed for the bar and were discussing the day's events and the situation that we now found ourselves in. Which was: we were smack in the middle of a war that had broken out before our very eyes.

A voice coming from the window confirmed our fears. "Jesus, look at that." We grabbed our drinks and shot outside and were now looking at gun tracers streaking up into the night sky. We all knew that we were now in a very serious and dangerous situation. We hadn't heard a thing from the American management. Where were the communiqués? There is usually a build-up to a war but there was nothing. We each grabbed a plate of sandwiches and headed back to one of our rooms and had, well, one or two beers. I think the whole lot of us got rather pissed that night. The next day I was woken up very abruptly at five-thirty by the sound of a jet screaming over the camp. I leapt out of bed, quickly dressed, splashed some water on my face and headed to my truck, followed by some of my crew.

As we drove off, I made it clear to them that I was no longer their boss and that I wasn't going to tell them to come with me to the plant. One guy suggested getting out at the main administration office to check out what was going on there and to get our passports, or at least find out where they were kept. We carried on towards the plant. It was very quiet and ghostly; machinery was either shut down, or shutting down. You could hear hissing and bangs and all kinds of weird noises. I opened the office door and told the men to wait for me and headed for the control room to see the operations manager. He was standing in the middle of the room in conversation. I waited, then all of a sudden a large explosion followed by another. All hell let loose, alarms were going off and the whole place shook, just like you see in the movies. Dust was coming from the ceiling, lights on the control panel were flashing and all kinds of mini explosions were coming from the plant. We were surrounded by some nasty and very volatile chemicals in the tanks all around us. I didn't wait any more. I headed back to my office and said, "Let's get the hell out of here."

We ran through the plant, expecting to be hit by another bomb, climbed into the truck and headed away from the scene, picking people up as we went along. The Iraqi authorities had not allowed us to drive on their open roads but I for one wouldn't be taking any notice of traffic policeman that day! Soon, we realised that we were heading in the wrong direction, towards Iran, which was not really a wise thing to do in our situation. We stopped to get our bearings and access the situation. Looking back at the camp, we noticed that it had taken a hit too – smoke and flames were coming from it. "Christ, there's women and kids in the camp." I turned the truck round and headed back. What we saw was devastation. There were dead bodies scattered, bits of bodies, and the swimming pool was red. The Lloyds surveyor, whom I'd been working with, had only just returned from leave. He had been living next door to me. The poor sod had taken a direct hit - there was nothing left of him. His colleague who lived next door to him was badly burned. My front door had been blown off, had gone through my bedroom and embedded itself in the bathroom wall. I had missed it by ten minutes.

It's amazing what happens to you in a situation like that: you are moving at normal speed and yet, things seem to slow down, as if your mind is in overdrive. We got back in the truck and started off for the Kuwait border. As we were passing one of the houses, a Geordie, who we had picked up, started shouting, "Stop, stop here, this is my house!" He jumped out and we waited for him, half expecting to be the next target. "What the fuck is he doing?" said Taffy, in his Welsh accent. "He may have a mate in there," someone said. "I wish the fuck he would hurry up, for Christ sake." More than a wee bit of patience was needed in the situation that had been thrust upon us, being a sitting duck for starters. Geordie came running out of his house carrying a case of Heineken under his arm! Well, what could we say? But it was one of the most welcome beers that I have ever had. We reached the Basra-Kuwait road and it was absolutely chaotic. The Italians had already decamped the night before. There were masses of people, all trying to get out at the same time; tempers flared and arguments started, as you can well imagine. The Kuwaitis wouldn't let us through. We were now sitting targets, on the border having to wait alongside Iraqi supplies and use one toilet. This, we had to endure from seven am until nine o'clock in the evening.

Stories were filtering through. We heard that a Brit had commandeered a bus and was ferrying people from the camp to the border, another was picking up people on his Honda 50 moped. Then we started to hear of casualties: my American friend, who had invited me off the street and into his party when I was lost on my first weekend at the camp and couldn't find my house, had been killed by the first bomb. He had been in the garage getting his car serviced. The garage was less than a hundred metres distance from the control room that I was in at the time. The Iranian Phantom pilot would no doubt have been targeting the control room which is the heart and brains of the plant and missed by a micro-second, less than a blink of an eyelid at the speed of a Phantom. Hence my Phantom story at the beginning of this chapter. (Little did I know, I was eventually to meet this very pilot later in Iran...) The Scottish construction manager who was also in the garage at the same time bought it too, poor sod. My American friend's wife and young daughter were frantically asking everybody if they had seen him. I didn't have the balls to tell her the tragic news and I regret to this day not doing so. But what could I have done or said to them to make matters any better?

War is so horrid and unpredictable. You just can't plan the next move in such ominous circumstances. Finally, the Kuwaitis let us through. I asked an Iraqi soldier if he had a vehicle, he shook his head. I pointed and said, "Well, now you are the owner of that a new American four-wheel-drive truck." I gave him the keys and walked through 'no-man's-land' to the next border post. We were driven to the British Embassy, where the staff greeted us warmly and served us drinks. Sitting round the embassy swimming pool, drinking cold beer, we related our tales. That night, we were put up at various hotels and must have looked a right ragtag and bobtail lot. I reached a phone and spoke to my wife and she told me that one of the Brits had got back to the UK and had told the story of what had happened on the BBC news. She telephoned my company for an update and they hadn't heard a thing and didn't know, yet people in their employ had been killed and injured. My daughter's teacher passed the message to her, as she was sitting in a math's class, that her dad was alive and safe. She burst into tears.

I spent the night at the Marriot Marina hotel, a ship converted into a hotel. Just the place for an old sea dog. We all spent the next day round the swimming pool, waiting for news, and in the evening, we convened in the hotel dining room and were told to make our way to the airport! At the airport, we were informed by KLM of the seating order. They had sent out three of their 747s to evacuate us. The women and children were obviously first, the remainder waited. There were still some seats left on the first flight, so we drew lots and I was one of the lucky ones. Our intrepid American management finally arrived. "OK, you guys, this is what we are going to do," said one manager. They were told in a very Scouse, Jock and Geordie diplomatic way what they could do with themselves. They had not kept us informed in any way of what was brewing. They had shut the plant down, again without giving us any information; we had to get our own passports and make our way to the border and here, in the safety of Kuwait airport, they had the audacity to come out with their John Wayne bullshit.

Feelings were now running high. On board the plane, I was sat in-between a Geordie and a Jock. KLM were brilliant and they soon made sure that everybody had a drink in his/her hand. I had feelings of euphoria, sorrow, anger and bewilderment. I think it is now called 'post-traumatic stress disorder' and people have received large amounts of money in damages. We didn't get a penny, or even an apology from two of the world's biggest construction companies, who had kept us there and must have known, or at least had an idea, of what was going on. Soon a song broke out and some laughter. An American guy stood up at the front to give us a lecture and asked us all to be quiet. He was either very brave or stupid. I would have said the latter as just about everybody wanted to have a talk with him. I don't know how I managed to constrain the Jock and the Geordie from having the first conversation with him, as we were the nearest and they were both straining at their leashes. Just before we landed, the captain came on the intercom to give us information, which made a pleasant change. He told us that the media would be at the airport and to be careful what we said to them. I had a microphone thrust in front of me as soon as I got to the bottom of the plane steps and, remembering what I had just experienced – people blown up, my American friend killed, his wife and children at the border - I spoke my mind and said awful things about the Iranians. At the hotel, we assembled and were told that the project had now been abandoned, which I thought was quite obvious. They handed out Dutch guilders, to enjoy ourselves on the town? Just what we felt like after the ordeal that we'd been through.

It was four o'clock in the morning. After a few drinks, we walked morosely through the cold wet cobbled streets of Amsterdam, looking in red shop windows, the ones with the beckoning women in. We were all suffering from a terrible anti-climax and were at our lowest of ebbs. The next day at the airport, a strange feeling came over me. I kept thinking that I was at Liverpool airport and actually felt like crying; delayed shock is a very strange thing. I arrived home and hugged my wife and kids and was having a cup of tea, telling them all about what had happened, when a car backfired as they did those days. I nearly jumped out of my skin. For years after, I kept having the same dream of jet planes coming for me and would wake up in a cold sweat.

We turned the six o'clock news on and the first thing that came on the box were war scenes from the Gulf, tanks rolling by etc, then it switched to the airport and I could see some of my colleagues from the project – some were in wheelchairs, some were bandaged and then my face came on the screen, being interviewed coming off the plane in Amsterdam. I had really torn into and criticised the Iranians for what they had done. Would I live to regret my outpouring of verbalism if

I ever went to Iran in the future? How strange, it wasn't repeated on the nine o'clock news. And yes I did get to go to Iran but that's another story!

Marsh Arab children (little girl in the red dress at the back).

SYRIA, 1ST TRIP
1981

(In which I dodge God and dress the Devil.)

'Culture shock.' Bingo! The magic words. I had hit the nail on the head. It was during an interview for a project which would eventually have me flying around the world in one trip. "You have to get used to the culture shock, in many places," I'd said. This seemed to perk up my future boss who had informed me that he was recovering from a tummy bug and who very much agreed, no doubt having had his fill of similar shocks himself. I felt somewhat chuffed at getting this contract as the agency told me that the shortlist for this job was two inches thick in paper. The weekend before I flew out, I was watching Brian Waldron MP's programme on TV about world affairs. This episode was all about Syria and the late President al-Assad. I was hearing things like 'harsh regime, torture' and seeing tanks rumbling along and all the usual nastier sides of life! Nevertheless, on the following Monday, I travelled to Cambridge where the company's main office was located. After being introduced to all the staff, I was given an office where I spent the remainder of the week perusing the blueprints and equipment and generally familiarising myself with the project, which was at the Faculties of Electrical and Mechanical Engineering at the University of Damascus.

I flew out with my now boss, who was called William and had recovered from his bug. During the flight, the stewardesses came round with the drinks. Not wanting to make a bad impression, I just ordered a beer. William ordered an orange juice which he drank in one go, then turned round and fell asleep. I quickly put an order in for a G&T and another beer before he woke up. A couple of orders later, William woke to tell me that he was feeling a bit sick. He had eaten something on the plane on a previous flight that hadn't quite agreed with him, so we checked in at the Sheraton and headed to our rooms. Despite his delicate stomach, William asked me to pop round to his room before dinner so he could brief me on tomorrow's meeting with the Dean. At the briefing, he asked me what I'd like to drink? 'Sod it,' I thought, "G&T please."

Very soon, and quite unexpectedly, the conversation drifted away from the university and onto religion. 'Jesus H. Christ,' was my first thought, and asked for another G&T. I have nothing against religion and people's beliefs, in fact one rector that I know (a relative), after graduating from London University as an electrical engineer, saw the light, switched it on and joined the church. All power to his elbow, watt! But this conversation was starting to bore me to tears, especially when he handed me a book on religion – not the Bible, more of a 'teach yourself religion' type thing - and I found myself asking awkward questions like: "If you died and your wife remarried, what would happen when she and her new husband entered the Pearly Gates? Isn't it going to be somewhat awkward when all three of you meet up?" I couldn't help it but had to bite my tongue so I didn't come out with any double entendres as this might push him over the edge. As it was, I could tell it was getting at him, as he poured himself a double orange juice on the rocks. I left the book in his room on purpose but he gave it to me over breakfast the next day. What does the 'H' in 'Jesus H. Christ' stand for?

After our meeting with the Dean, we were shown around the laboratories. I noticed that most of our equipment was still in its wooden packing cases. A certain 'Dr Freebie' was waiting for us in one of the laboratories; he and I would be working together. Freebie (not his real name but it might as well have been) turned out to be real pain in the arse as he was an opportunist, was definitely in it for the backhanders and loved going to restaurants, providing that you paid the bill! William and I returned to the hotel and spent hours going over the contract and various technical details as he wanted to leave the next day, 'thank God.' He would fly in one day and we would then spend all day and evening going through everything in a mad rush, so that he could fly out the next. Soon after he'd gone, I was having dinner in the hotel restaurant which incorporated a dance floor, stage and a disco. Without the restriction of William, I ordered a bottle of wine, followed by a cognac or two. Feeling quite mellow, and as the disco had started, I fancied a dance. There was a Syrian guy, sitting with two women, so I popped over to ask one for a dance. He opened his coat so that I could see his gun. "I take it that means no, then?" I didn't wait for a reply!

The following day, I had to locate our agent, Ismial, which took ages as none of the taxi drivers could find the place. There was only one thing for it: visit the commercial department of the British Embassy, they'd know, which they did, so now I had local support and someone I could get my allowances and expenses from. On my first working day at the university, I was met by Dr Freebie. As we walked to the laboratories, I noticed many bullet holes in the walls and spent cartridges on the ground. Dr Freebie saw my puzzled look. "Student riots," he explained.
"What, here? Was anybody hurt?"
"Only two."
"Badly?"
"Deadly. Dead!"
A sudden feeling of foreboding crept in. No wonder William wanted to leave on the next plane out! Brian Waldron

loomed up again with his lisp: "Harsh regimes, torture!" We arrived at the main laboratory where a Dr Nabih, who was later to be known as 'Dr Maybe' had been waiting for us. Dr Nabih's stock answer to every question was: "Maybe!" 'Nabih' in Arabic means 'smart' and he very much was... at dodging questions. So here we were, Dr Freebie and Dr Maybe and me.

Obviously, the first thing we had to do was to open the packing cases, identify the items and make sure that everything was there according to the packing lists and also check for any damage. This was the metallurgy laboratory. They also had a foundry lab, a stress lab, chemical lab and the main laboratory, which was where the electron microscope was to be situated. 'BANG, BANG, BOOM!' What the fuck was that? There was this lad of about fifteen years old, banging hell out of the packing cases with a huge sledgehammer! "What are you doing, you idiot?!" I was thinking about the electron microscope. He just grinned, showing all his bad teeth. We eventually unpacked all of the equipment and to my surprise, nothing was damaged. A miracle! Was William's book working? Not that I'd read any of it. We moved all the now empty packing cases outside to give us room. Needless to say, they evaporated overnight. The thief did me a favour by getting shut of it and they even took all the nails, bent ones as well!

I was getting awful stomach cramps every day soon after lunch – why was this? Then it dawned on me: the salads. We used to send Bad Teeth out to buy lunch for us and of course salad in these countries either isn't washed, or is washed in dodgy water. From then on, I brought packed lunches from the hotel – end of stomach cramps! Of course they wanted the electron microscope in a lab on the top floor, where else? This meant having to negotiate several flights of stairs that had a more than its share of tight bends. A motley crew arrived to move it, headed by the daft lad with the bad teeth. 'Oh Jesus Harold Christ.' It's amazing how often an atheist calls to God in these situations. For the next hour, my heart was in my mouth as it was manhandled. Every time it took a knock, I cringed. The mad lad would laugh and say my surname repeatedly, then they all laughed. I had to pinch myself. 'Was this really happening?' It was just like the movie 'One Flew Over the Cuckoo's Nest' where Jack Nicholson played the part of a normal guy incarcerated in a lunatic asylum, but this was for real.

Again, I checked the microscope: no damage, another miracle. Maybe I should read this book after all. The next problem was getting it plumbed in as it needed water in its cooling system. This meant hiring a plumbing contractor. I also needed to install a deioniser in the system with the water being brackish. Of course the Syrian plumber just didn't have a clue what a deioniser was. He had a point, where the hell do I get a deioniser in Damascus? The electron microscope's cooling system would soon choke up with all the salty crap and debris. It was lucky just to have water, let alone a deioniser. But by a pure miracle again, I found one! William's book, again?! To a diehard atheist, this was getting embarrassing. To arrange for the electron microscope engineer to fly out, my timing had to be spot on. First, he had to be available, then his flights and hotel had to be booked and visas obtained. I had given the go ahead for him to fly out on the Sunday so he could make a start first thing Monday morning. The plumbing contractor had fallen behind schedule and it was now Thursday morning and they were off on the Friday (their holy day) and the weekend was a local holiday. I noticed they were now starting to wind down and by the afternoon, they started to pack up to go.

"We come back Tuesday."

"Tuesday! You only have an hour's work to do, please can you finish it?" I had visions of the engineer arriving Sunday and checking into the Sheraton, which is not the cheapest of hotels. The company were being billed £1000 day for his services and he would not be able to start until Wednesday! "Please can you finish before you go home?"

"La, la, no, no, we come back, finish Tuesday!"

I had no other alternative. I locked the bastards in and told them to 'finish, or they no come out until fucking Tuesday.' Amazing what a bit of tact will do!

One day, Dr Maybe said, "Why you take taxi? Come on bus with me." 'Great', I thought. I didn't fancy that but didn't want to hurt his feelings. We no sooner set off on the crowded bus when a heated altercation broke out at the back between two irate males. As we were all packed in like sardines, there was no way to get out of their way or get off. Then to my horror, one of the combatants pulled out a gun and stuck it up the other chap's nose, as you do! Was I dreaming all this? The lisping voice: "harsh regimes, torture, murder," went through my head again! I closed my eyes and wished that I was dreaming. The tension and stress was slowly draining me, especially through lack of sufficient sleep due to the noise outside the hotel, as most of the taxis were minus exhausts and had their horns attached to their brake and accelerator pedals and this would go on all night long. I somehow managed to avoid being shot at on the bus. Can you imagine?

"How did he die?"

"Oh, he was shot on the number six bus, on his way home."

Hardly James Bond stuff.

Back at the hotel, I had had a hot bath, followed by a couple of stiff drinks, and was beginning to wind down when the telephone pulled me out of my bliss.

"Hello. You must come quickly."

"Eh?"

"Quickly, you must come."

"Who the hell are you and why and where must I come quickly?"

"I am customs officer and you must come to airport immediately."

The key in my back started to strain as it turned. I arrived at the airport and noticed that the whole area had been cordoned off and the police were everywhere. What the hell is going on here? I introduced myself to the police, who were not over-friendly.

"Is that your box?"

"Which box?" as there were dozens. He pointed with his handgun. There must have been about ten policemen all with their guns at the ready. And I'd forgotten my book! The box in question was from the company that supplied the electron microscope. It had been opened. I slowly lifted the lid off the packing case and heard the safety catches coming off the guns. You know those movies where they are defusing a bomb and the perspiration is dripping off their faces. Well, I now know how they feel! I knew it couldn't have been anything dangerous and that it must be the part I was expecting but it was the guns behind me that I was worried about. It turned out to be the optics for the electron microscope and they were in a metal container immersed in liquid nitrogen. I took the lid off and immediately the liquid nitrogen gave off a vapour. I heard gasps from behind. The customs and police had thought it was a bomb, which was fair enough as it did look dodgy. And it was vaping too. After all the excitement, I returned to my hotel, feeling like jelly.

The bean counters from finance in the head office were now on my back to get our acceptance certificates signed, otherwise we all starved. The certificate had to be signed by seven doctors and in order to achieve this, I had to first prove that everything was there and not damaged, then demonstrate, test and prove that the equipment worked and did what it was supposed to do. The tricky part was trying to get seven doctors from the university together at the same time, go through everything with them and answer any questions they may have. Then they all had to agree to accept the equipment, then the magnificent seven all had to sign at the same time. The signed acceptance certificates would be sent to the UK, go straight to the bank and the company would be paid. Simple! But you are dealing with Syria and have you ever tried to get seven busy university lecturers assembled at the same time? It would have been easier to get Snow White and the Seven Dwarfs. Time was running out and people were getting impatient, as penalties were looming ahead. 'Bribery,' that was the word. I would bribe them! So, with the promise of a slap-up meal and a visit to a night club: "Another cognac, gentleman? No problem. Oh, and can just sign these bits of paper?" I had them all signed in one go. Who says some people are corrupt? Nonsense, they just like to wine and dine their customers!

There were two watering holes in Damascus that I frequented. One was The Pig and Whistle in the British Embassy and the other was the Marine Bar, the American equivalent. So it was Sunday lunch and Wednesday evening at The Pig and Whistle and Friday evening at the Marine Bar. One Friday, I found myself in conversation with some teachers from the American School about the theatre. They were putting on a show with their students and it wasn't going well. Back in the mid sixties, I had a part-time job at the Royal Court Theatre in Liverpool on lighting and sound, and my claim to fame was meeting, at the tender age of seventeen, the movie star Joan Collins (pre-Dynasty) and her sister Jackie Collins, the writer. A musical was on, 'The Roar of the Greasepaint - The Smell of the Crowd,' for a two-week run and it was co-written by Anthony Newley and Leslie Bricusse. Newley at the time was married to Joan Collins. It was a Monday evening and we were into the second week. I walked through the stage door and virtually bumped into these two gorgeous women whom I didn't recognise. Anthony Newley saw me and said, "What the hell are you wearing?" I had on a pair of white trousers and a pale blue shirt. As I wasn't wearing a dress, I wondered what he was on about.

"Have you seen the set and the changes?"
"No, I've only just arrived."
"Pop your head in and have a look."

I looked at the stage, came back out and said, "Some clown has removed all the curtains?"

"That's the new set and the clown is me."

Oops. He'd had all the back and side curtains removed, which meant that the audience would now be able to see all the backstage junk, ladders, bits of old scenery, fire buckets, the lot! It soon sank in that the audience would also be able to see me. During one of the scenes, a young actress called Dilys Watling would stand on a box wearing a flimsy raggedy dress and I had to stand behind a side curtain (that was once there) holding an electric fan, and blow her dress. That was embarrassing enough for a seventeen-year-old, but now the audience would be able to see me holding the fan. Newley thought it was brilliant. He was way ahead of his time, with his 'Stop the World - I Want to Get Off' and 'The Strange World of Gurney Slade' that was on the TV at the time. He liked the idea that I would be seen and asked me to turn up every night wearing light-coloured clothing for the remainder of the show. There's always a joker in the pack, always a cardboard clown! I will always remember walking down Lime Street with him, trying to find a late-night chemist that was still open to get some aspirin as, yes, JC had a headache.

Back in Damascus: "We only have three weeks to opening night and I can't do a thing with them!" came the cry for help from one of the teachers at the American School. I agreed to assist her as best I could as I didn't have anything to do in the evenings anyway and was looking forward to it. The following

Monday, I went along to the American School to observe and see how best I could help out. The cast were at that awkward age, early teens. They didn't know their lines and kept arsing around. The costumes were dreadful. The teacher, who was also the director, had one of the main characters, Lucifer, dressed up with a tail on his costume. Of course he felt ridiculous, especially with his peer group watching and taking the piss. I went over to him and asked him if he had a leather jacket, white T-shirt, jeans and a pair of cowboy boots. Being an American, of course he did. "Oh, and put some grease on your hair too." The following night Lucifer showed up looking like the Fonz from 'Happy Days'. I handed him a whisky bottle, half full of cold tea, and a dummy cigarette. He was now the modern-day devil! And what a transformation: now he looked cool to his peer group and really gave it all he had.

We couldn't find a top hat for the undertaker so I ended up making one out of an empty paint tin and cardboard. We dressed him in a black jacket and pinstriped trousers, and now he looked the part too. I asked the other kids what they thought they should be wearing. After rummaging through their parents' wardrobes, they too turned up dressed the part, as they'd really had some good ideas and their parents had surprisingly good wardrobes. Now they had confidence in themselves and didn't feel stupid or awkward in their costumes. It also made them want to do it. I mentioned that we may use a prompt and would have him/her hidden behind the sofa and if anybody fluffed their lines, the prompt would jump up and assist. It was a farce, after all. This did not appeal to them one bit, but it did the job in making them learn their lines... well, almost. On several occasions they had to ad-lib and improvise, which seemed to make it even funnier. The opening night came and I could see big, expensive, chauffeur-driven cars drawing up outside and well-dressed people sitting down in the audience.

"Who are they?"

"Oh, didn't I tell you? These kids are ambassadors' children."

"Gulp." That explained the wardrobes.

The show went down a treat and the audience just loved it. I ended up being coaxed out to give a speech of a kind, which was very ad-lib. I would never have made an actor; it's much too scary in front of all those people.

I had to change hotels over the weekend as I couldn't take any more of the taxies minus exhausts etc. It was a Saturday night and the hotel receptionist recommended a bar called the Crazy Horse and I set off. Nobody seemed to know where it was so I took a taxi. Some thirty minutes later I arrived, and the taxi fare wasn't cheap! Still, I was there. Some time later, I headed back to the hotel and walked along some fifty yards, trying to get a taxi for the long journey home when I came upon my hotel! The bastard had driven me round in bloody circles to bump up his fare! Lesson learned. I was up early the next day as Dr Freebie had arranged to take me to see the ruins of Palmyra, which was Syria's equivalent to Athens. I was quite amazed as I'd never heard of this place before. Now the so-called Isis, Islamic State, had taken it over and were ruining the ruins, so to speak. We were driving along the famous Damascus road, when we were overtaken by a fur hat, obviously a Russian, as they can be found in these unusual places dotted around the world, wearing just such hats. Now what the hell was he up to? No good I am sure.

My work was completed, all the necessary bits of paper were signed, reports done and I headed to the airport. It is always a wonderful feeling driving along the airport road on your way home - until we reached the road block. What the hell is up now? A plane had been hijacked and hostages taken. No wonder they call the Middle East the tinder box of the world. I had to return to Damascus until it all ended. I checked back into the same hotel where a fax was waiting for me: 'Can you please go to Athens on your way home? They have a problem at the University of Crete with some of the equipment that we have sold to them. Head for Athens and see our agent there, his name is Candelas.' The following day, I rescheduled my flights and headed once more to the airport. The hijackers had killed one of the hostages, a young American sailor, and had thrown him out onto the tarmac to show that they meant business. What callous bastards. Then they blew up the plane before President al-Assad could intervene.

I arrived in Athens where I could wind down. Candelas picked me up from the hotel and took me back to his office. The problem was that the Agricultural University had purchased a plough and couldn't assemble it.

"Now where do I come in? I'm not a ploughman," I found myself saying.

"Just pop over there tomorrow and see what you can do. I have booked your air tickets and you will be assisted by my accountant."

"An accountant?"

"We don't have anybody else."

"Well, I suppose he can help me with my expenses."

We arrived at the university and proceeded to erect the plough, which was in bits on the floor. Amazingly, the accountant knew a bit about ploughs, so he did come in handy after all. We toiled all day and completed the job, only to find that we had missed our flight and had to stay overnight. Crete is sweet, Damascus is Dam-crazy, I thought, so no problem there! I received another message from Cambridge. They were having problems at the university in Damascus with the electron microscope, so a quick flight back, problem solved.

I then had to get back to Athens for a meeting with our agent and had arranged to meet my wife there for a holiday. The travel agent in Damascus could only book me a flight to Athens with a stopover in Beirut which was still in the throes of a civil war. For some reason, probably not helped by all the hectic criss-crossing travel, it never crossed my mind that the hostages Brian Keenan, John McCarthy and Terry Waite, the Archbishop of Canterbury's special envoy, were all being held captive there. The plane landed in Beirut and I got off to stretch my legs as I had an hour before take-off. In the airport lounge bar, I couldn't help notice that most of the walls had, I assumed, been strafed by bullets which did nothing for the ambience of the place but a beer made me forget about it. Well, no it didn't! So another was ordered in quick succession.

The flight was called and I made my way to the gate. I handed my boarding card and passport to security, when two scruffy soldiers/militia, armed with AK47s, checked my passport and beckoned me to follow them. I didn't know know what was going on and was feeling somewhat apprehensive to say the least, now I'd remembered the three hostages. They had nothing to gain by holding me, I may soon have to tell them in broken/pidgin Arabic, but moment later, I was being led out onto the air strip. What a wonderful sight the plane was outside on the tarmac. My escorts escorted me to the waiting plane and I assumed, hopefully correctly, that they were there to stop me being taken as a hostage by some faction or other. I counted: hundred, fifty, twenty, ten yards to go, then a flight of steps. A sigh of relief was heard throughout the airport! Once those wheels lifted, a large Scotch was my first and second order, then it was peaceful Athens to smash some plates at a fireplace!

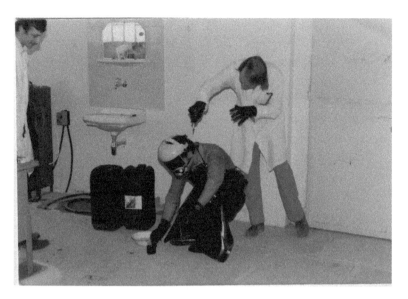

Adjusting an instructor's brain at the University of Damascus.

Me in Palmyra, before Isis ruined the ruins. Photo by Dr Freebie.

Ecuador
1981

(In which I dunk a Globetrotter and trot the globe.)

On the flight from Korea en route to Los Angeles, I was sat next to a fat, follicularly challenged Korean who took up most of the space and stuffed his face all the way to gaining an extra two stone before we landed. Some years later, a similar Korean Air 747 was shot down by the Russians on the same route as this flight. The Korean pilot had somehow lost his way and wandered into their airspace – this panicked the Russians enough for them to blow it out of the sky. They found it absolutely necessary to shoot down what was obviously a passenger airline, a massive Jumbo jet, with the loss of all on board.

Getting through passport control was about as easy as getting through the Berlin Wall before it was knocked down. I mean, some of the questions you are asked by the American immigration… I had a couple of hours before my connecting flight to Quito, the capital of Ecuador, so the bar made an excellent waiting room. I hadn't heard my name being called out by the Korean Air ground staff and had forgotten to change the time on my watch. A Korean Air official approached me.
"You Mr Hopley?"
"Yes."
"Well, you better move your ass, the gate is about to close."

I ran like hell and only just made it. Imagine trying to do that now in the States! I sat down and couldn't believe who was in the next seat, the fat Korean. F*ck. I later found out that they thought we were travelling together. The airline was Ecuatoriana and the plane was painted psychedelically like John Lennon's Rolls-Royce. I could well imagine that it had changed ownership once or thrice, one being Ugandan Airways. It was worse than when British Airways changed from that regal look to that dreadful colour scheme where the tails looked as if they were painted by drunken chimpanzees.

I tried to order a drink but the stewardess couldn't understand my feeble attempt at Spanish. "Por favour, uno el gino un tonico mucha gusto." Nor could she understand my gestured actions either. Fatso sniggered and stuffed another whole chicken into his void. I went to the galley to point out what I wanted. An Ecuadorian came to my rescue. She had seen and heard the appalling attempt by Juan Dope.
"Would your friend like a drink too?"
"That isn't my friend, just a fat Korean who happens to sit next to me wherever I go."
"You come sit next to me and my daughter, we talk." She had been to see her in-laws. The next moment, the plane took a dip. I was in a psychedelic airplane flying over a South American rainforest on a magical mystery tour. After all my trouble getting it, my G&T ended up on the ceiling. I was beginning to feel a wee bit of trepidation creeping in and I couldn't get another drink as the bloody seat-belt sign was on!

I had more problems when we landed, this time with immigration over my visa, and they were going to put me back on the same plane to go back again. Eventually, after a lot of arguing and bartering, the agent finally got me through. By the time I reached my hotel, I was completely worn out with jet lag and the aggravation and was soon unconscious. The phone rang and roused me out of a deep sleep. It was some crazy female who couldn't speak English and never stopped giggling. This was just what I needed after flying all night in a psychedelic airplane without sleep, a bloody psychopath on the phone. I was later to find out that she was the maid of the women that I'd met on the plane, inviting me to dinner. They were having their little joke. I didn't laugh!

The project in Quito was at an army university, where we had to install, pre-commission and commission equipment in the thermodynamics, heat engines, steam and fluids laboratories. The first problem that surfaced was a logistical one. All the equipment had been stored in a warehouse on the edge of a rainforest about twenty miles from the university. The second problem was that I had to organise and motivate a battalion of raw army conscripts to load scientific equipment onto trucks, then transport them over rough terrain, unload the crates in the university car park and transfer the equipment to the laboratories. This was not going to be a walk in the park! To make matters even worse, my boss, William, had flown in, with his usual diarrheic sickliness, and was now insisting on assisting. He climbed onto the fork lift, immediately crashed into the army truck and started to lift crates of very expensive and delicate scientific equipment. "William, I hope you realise that we are not insured for you to drive a forklift truck!" He soon got the message and left it to the conscripts to wreck their equipment. At least if they ballsed it up, we'd be covered.

After the struggle and trying to make sure William the Arse-istant did not assist any further, I finally got the equipment to the university in one piece. For the first two weeks, I was assisted by Andy, a Taff who used to work for Racal the electronics company and now worked for our Ecuadorian agent. He had married a local girl, had a baby and made Quito his new valley. As well as Andy, we had hired a bunch of local labourers – one was smelly and I'll come to him later on. The project was going very well and I really enjoyed the challenge.

The hotel was spot on, as a four star should be. Quito had some wonderful restaurants and bars and one bar in particular was owned by a portly Austrian, Gerhard. Gerhard was quite a character but one night he nearly got himself into a tribulation. The bar was full of Americans, mainly servicemen. I used to smoke at the time, and shouted over to Gerhard, "Twenty fags please, Gerhard." A fag in Liverpool is a slang word for a cigarette; in America a fag means a gay person. I didn't half get some queer looks, excuse the pun.

No sooner had the commotion died down, when several black servicemen entered the bar, each built like Muhammad Ali. Gerhard forgot himself and asked one of them what he wanted, but in a Southern American Uncle Tom accent. You could hear a pin drop on an Axminster carpet on the Moon. I thought, 'Oh shit'. The American GI gave poor Gerhard the eye. Gerhard had that hare-in-car-headlights look. A cry of "Where's me bloody fags, Gerhard? I'm gasping!" changed the situation to laughter. Good old Scouse humour saved the day and Gerhard's life continued! He couldn't wait to buy a round of drinks for everyone. I even got my fags, the ones that you smoke. On another night at Gerhard's, it was after closing time and Andy and I were discussing measurements of the engineering kind, it escapes me now as to why. Gerhard chipped in: "How wide do you think this room is?" We all gave our answers, I was the nearest. He got out a tape measure. Then it started and soon we were betting for money. "How long is the bar? How long is a fag, sorry, cigarette? How long is the whatever?" And this went on for about an hour, as we merrily got smashed. With me being an engineer and used to measuring and measurements, Andy and Gerhard didn't stand a chance.

One of the main headaches of the project, as usual, was to get the equipment handed over and signed off so that our company could get paid. To do this, first we had to check every component and spare part for shortages and breakages, then show it to whoever was the signatory, as it had to tally with the paperwork. It could be most difficult trying to explain what a certain part was and to match it to the lists - until Edwin came along. We couldn't quite fathom out what Edwin's function was, only that he must have been well-connected to the university's hierarchy as he was too old to be a student and too dim to be a lecturer. By a sheer stroke of luck, I happened to tell him that he looked like Jack Nicholson. He did in a funny sort of way. Edwin used to wear these ridiculous dark sunglasses similar to the ones Jack Nicholson wears, but he's a movie star! I had said the magic words. So I continued to butter him up with "Hi, Jack!" and he would sign for anything. So that shaved some time off our schedule.

More troops arrived on the project: Tony and Arthur, two Geordies. Tony was the principal lecturer and Arthur was a senior lecturer from Newcastle University. They used to work for our company during the semesters to commission some of the equipment. Then there was Alex the Jock. Alex was the mechanic with the ubiquitous cigarette in mouth and Daily Mirror in coverall pocket. That bloody cigarette nearly brought my time on this planet to an abrupt end. I was in the middle of leak-testing a bank of a dozen propane-gas bottles that had been connected together with joints, piping and elbows. The propane gas was used to fuel a ramjet engine. The bottles were five foot in height and one foot in diameter and I was in the middle of them, when Alex, cigarette in mouth, popped his head through the door. I politely told him to "F***ing f*** off! / go away!"

Tony and Arthur were real characters, both very clever and extremely witty too and it was a joy to work with them. They both acted like naughty little boys let loose in a playschool with lots of toys to play with. Some toys! A mini rocket engine, a ramjet engine, gas and steam turbines for starters.

Both of them were fully paid-up members of, funnily enough, the Boilermakers Union, which I couldn't quite understand, with them both being lecturers in mechanical engineering. Both loved football and this is where Smelly comes in. During our lunch break, we would have a knockabout in the yard outside. Some of the local labourers were watching and we invited them to join in, one being Smelly. He wore Wellington boots all the time; I think he probably slept in them as well. You could always tell when Smelly was around, you wouldn't see him but by God, you could smell him. You would be doing something in one of the labs and this nauseating pong would drift in. It was Smelly in his Wellies. "Buenos dias, Smelly," you said in haste as you dashed out the door for some fresh air. Tony and Arthur would walk in and say in unison, "Smelly's been."

The weekends were good fun, we were sometimes invited to barbecues up in the mountains with the university staff. A pig would be roasted on a spit, with lots of my favourite vegetable, corn on the cob, and many local dishes, accompanied with lots of beer and wine. Another day out was a trip to a small town just next door to Ambato in the Andes. You'd have to be up very early in the morning to catch a very rickety bus from the local marketplace. It would take you through mountain passes and along dirt roads. If you looked out of the window and were on the wrong side of the bus - I lay emphasis on 'wrong side' because it was a constant worry seeing the bus's wheels only inches away from the edge of the road and certain death. It was a fantastic view - if you dared to keep your eyes open. You would pass through many small villages that were probably not even on the map. The town reminded me of one of these towns in a Western movie - all wooden stores, restaurants, bars and sidewalks, where you could buy very cheap, well made leather goods and a variety of well made souvenirs. Then the return journey in the dark and the thought of those wheels!

Some weekends you would spend around the swimming pool chatting and having a drink(s). One Saturday, I had been by the pool all afternoon, when this huge black guy came strolling along. The Harlem Globetrotters were in town and he was one of them. I, being a bit pissed, walked over to him with a daft grin on my face. He immediately grasped what my intensions were and started to run. With Quito being way above sea level, it took you a week or so to acclimatise and he had only just arrived so I soon caught up with him and started to drag him towards the pool. "No, man! No, man!" he was saying. Now this guy was at least seven foot twelve and me five nine. He could have knocked me over, no problem. We reached the pool. By now, guests and staff were watching the commotion from their balconies and windows. I pushed him, he grabbed hold of me and we both went in, fully clothed. We climbed out laughing our heads off and walked straight to the bar inside the hotel dripping wet! Followed by a great time. What a character and what a sport. I wonder how many people have ever pushed a fully clothed Harlem Globetrotter into a hotel swimming pool. Is that a claim to fame or what?

It had taken Tony and me weeks to set up the rocket engine, a scaled-down version obviously but it still did the business! First, I had to find a machine shop that could make a manifold - in Quito! Then go over the drawings in broken Spanish to explain exactly what was required. No easy task. We had just connected the oxygen and nitrogen pipelines to the manifold and all that was required was to screw in a plug. This was more intricate than it sounds, as it had a special fine thread. Tony, who never swore, spotted William my boss, sitting on a workbench, doodling. "Look at that useless c*nt," he said. William had seen us looking at him and immediately sprang up, dashed over to us and said, "What can I do?" Before Tony could get the second word 'off' out, William had already started to screw in the plug. To our horror he cross-threaded it! As the manifold had to withstand a high pressure, it was now ruined: we couldn't have taken a chance with it. The manifold had to be dismantled and a new one made. This set our programme back a week. We were now under a lot of pressure as the President of Ecuador was booked to inaugurate the laboratories, and to make matters worse, due to the delay, Tony and Arthur had to return to the the UK. We had ignited the rocket engine successfully, unfortunately we didn't have enough time to start the ramjet engine. The only thing for it was for Tony to go over the start-up procedures with me as I had to do it. Tony's last words were, "Bring a spare pair of underpants with you." Point taken!

The big day came. I couldn't eat much for breakfast. I opened the lab door and started the pre-start checks on the jet engine and consul and went outside to open all valves on the propane-gas bottles. I had the feeling of, I would imagine, a condemned man prior to the long drop! I went to the control consul and set the pressures accordingly and pressed the fire button. Nothing happened. I tried again. 'Oh shit.' I had enough pressure. I rechecked the igniters, none were working. A wire had somehow worked itself loose. Having rectified the problem, I went for ignition again – my mouth was dry and I couldn't swallow. Pressures spot on. It burst into life. 'Wow.' Being so close to a ramjet engine, even though I was wearing ear protectors, the noise was unbelievable! The whole ground vibrated and the sound was awesome. To have control over this fire-breathing monster was terrifying but I was enjoying it.

Fear is a strange thing. People have their own bogeyman and yet still enjoy the thrill. I used to scuba-dive and fly and both used to give me palpitations. With diving, you are sat in a boat, it's a grey overcast day in the Irish Sea and you are heading out. The sea is cold, dark and not inviting whatsoever and you are going down into that, twenty-five to thirty-five metres. It was a very daunting feeling. You go over the side and descend down the shotline into the blackness.

You immediately feel the cold as your diving suit starts to compress with the pressure. You reach for the air valve on your suit and crack it open to put some air into it, which warms you a bit. You had to be careful not to put too much in and not let it migrate to your feet otherwise you would shoot to the surface upside down like a Polaris missile and your lungs would explode and you would die (or get the bends and die!). Both are not the ideal thing for extending one's life. Your suit is still pinching... a little more air... now your mask is crushing your face, you breathe out through your nose to avoid mask squeeze, then pinch your nose to equalise the pressure in your ears. You are still going down. Without looking at your depth gauge, you know that you are deep as your mask now starts feeling loose. It gets darker and there may be basking sharks in the area that could be the size of a whale; harmless plankton-eating creatures, swimming around with their huge mouths open, so it would be slightly inconvenient to swim into one, literally. Flying, well, I won't go into it but I don't like heights, a perfect accompaniment. My two favourite examples of being scared and enjoying it!

Back to me igniting the ramjet engine. The next moment, everybody came running into the lab with worried faces. They'd thought it was an earthquake. The feeling of relief and a certain complacency flooded over me: I had conquered the ogre. On the day of the inauguration, the President arrived and we were all introduced to him. I gave a little welcoming speech to him and his generals; it was short and sweet. "Welcome, El Presidente, hope you enjoy your visit and have a nice day, por favor." Kate, our South American area manager, even though she was fluent in Spanish, froze in front of him and gave up halfway. He, of course, was into the second half of his term in office, which I'm guessing was about two months, with it being a typical banana republic. A presidential term is like a life of a butterfly and they are considered lucky if they don't get shot. This one had been told to relinquish his little empire rather expeditiously. On the whole, the day went well: I had not killed anybody while demonstrating the equipment and had kept Alex and his bloody cigarettes out of harm's way but it was a bit nerve-racking starting the jet engine in front of him and his generals.

So, speeches given, tape cut and it was time to party in Gerhard's. That night, Andy, Alex and I were invited to our agent, Nelson's, for drinks. Of course Nelson was a multi-millionaire, what else? We had stayed longer than intended. I wanted to pack as we were leaving the next day, so woke up in a blind panic. I had overslept. I just flung everything in my suitcase, no shave, no shower and no breakfast! What a way to travel. We got to the airport with barely enough time to catch the Psycho Airways flight to Miami, then London. I had flown round the world in one trip: Liverpool, London, Dubai, Hong Kong, Taiwan, Korea, Los Angeles, Mexico City, Ecuador, Miami, London and Liverpool. Michael Palin, eat your heart out. 'Pole to Pole'? This was 'Bar to Bar'.

Me in my lab coat, pointer in pocket, waiting for the president at the inauguration, Ecuador Army University.

KOREA, 1ST TRIP
1981-1990

(A troublesome stargazer and one big Bic-plic.)

I was standing in for the company's projects engineer who was away on sick leave and packing my case for the trip to South America, when the phone rang. It was my boss, William. "Can you make a minor detour on your way to Ecuador?"
"No problem, where to? ...Excuse me, did you say Korea?"
"Yes. We've received an urgent call from the University of Seoul – it's their dome."
"Their what?"
"Their observation dome."

Apparently the problem was, when the boffins at the university's stargazing department, the Faculty of Astronomy, were homing in on a celestial body, they would overshoot the target and go too far. Instead of being able to inch back, they would have to rotate the dome 360 degrees and if they missed again, it would be a pain in their 'Uranus' and rather time-consuming. As the dome was still under warrantee, we had to fix it. I had visions of a Korean Patrick Moore jumping up and down.

On the flight to Seoul, I mulled over what the problem could be. Well, I didn't mull too much, as I ordered a large G&T. There was a stopover in Hong Kong and Kai Tak Airport had a reputation for not being the safest of airports to land at, plus we were landing in a storm. I'd had a chat with a gorgeous

Australian female at a bar at Heathrow airport prior to boarding the flight but drew the short straw and got to sit next to a stern Israeli who 'tut-tutted' every time I ordered a drink. The plane was now buffeting all over the place. The aircraft was a DC-10, which at the time were going through a bad spell of various mishaps and were inclined to greet the land in a way they ought not to have done! I was reading the newspaper upside down (not intentionally, it's called nervousness) and to make matters worse, the Israeli guy was praying. When I was home, just before my flight, I had picked up a Blue Peter book of my daughter's. There was a chapter about Hong Kong and Kia Tak Airport's reputation, which said that take-off and landing were dangerous as you actually fly between mountains and skyscrapers. On the final approach, you can see all the little sweatshops churning out Barbie dolls and other plastic junk which makes its way to our shores and ends up clogging the beaches and seas, making someone loads of money. I could swear there were tyre marks on the roofs as we were that close.

William had done me proud as he'd booked me into a super hotel that provided a Mercedes to pick you up. I shared it with the Australian woman who was staying at the same hotel. She was meeting her friends, who were coming from Australia for a holiday, in Honkers. I bumped into them later in the hotel bar and we ventured outside to find a restaurant. The storm was really on us and we must have been mad as we were being blown all over the place, together with sheets of tarpaulin and corrugated steel from building sites. We passed stalls that were still open and selling various dishes of 'I'm not quite sure of what,' dead or alive. I remember the comedian Jasper Carrott telling the story of his trip to Hong Kong, where he was also walking past these fine-fare establishments, when a chicken escaped, darting under tables and chairs, followed by an irate Chinaman wielding a big chopper. I'm sure if you stood still too long in some of these places, you'd get the chop.

Wise chicken. 'Sod this,' we thought, 'let's go!' The storm was not abating and the rain was getting too heavy for us to go any further so we decided to dine at the next restaurant that seemed reasonable. Which just happened to be a Chinese restaurant. I kept thinking of absconding chickens. "What you want?" "Anything but chicken, please."

The next day, I boarded a Cathy Pacific flight to Seoul. It was still bucketing down - it sure rains in Hong Kong. We were soon above the clouds and away from the storm. I looked out of the window on China and suddenly felt all awestruck. Little ole me, flying over China. Wow. A feeling I would not forget. We soon touched down in Taiwan to refuel and I did the usual walk around the duty-free shops that sell junk you don't need but you somehow still gawp at and buy. I have so many locks, I could go into competition with Group 4 and also open my own off-licence. I can never catch up on the books I've bought either, so I don't know why I go in. I got chatting with some of the shop people and they all told me that their allegiance was with China. How strange. Chiang Kai-shek would turn in his grave. We were off again and soon dropping the wheels for Seoul. I found myself marooned at the hotel on the first night in Seoul due to a curfew that had been imposed on the people. They had to be off the streets by ten o'clock. You could smell the tear gas wafting throughout the hotel. The students had been demonstrating over something, I wasn't quite sure what. So I didn't venture outside to find out but headed to the bar instead.

The next day, I took a taxi to the university where I was met by one of the professors. He explained what problems they were encountering with their malfunctioning dome: it was not only annoying but it wasted their valuable time. I soon found out what the problem was - one of the limit switches had worked itself loose. This was soon rectified and they could now stargaze in peace. I couldn't help notice that the observatory was wet - in fact, it had had a good soaking. "I can see that you're having problems with your roof as well?"

"Every time it rains, we get very wet," the professor exclaimed.

I popped up onto the roof to investigate. The roof was flat and the drainpipe grids were blocked with leaves. The roof was like a swimming pool and when it rained, the water had nowhere else to go but through the roof and into the observatory. I took them up onto the roof to see for themselves and pointed this out to them. Amazing, they could find a star light years away and yet couldn't fathom out why their roof leaked. One would have thought I'd discovered a new planet the way they thanked me.

As I still had the weekend in Seoul before flying to Ecuador, the professors suggested I should do some sightseeing and that they'd send a car to my hotel to pick me up. The following day after breakfast, I decided to wait outside the hotel for the car. I was not there long when a big black limousine with a Union Jack flying on the bonnet drew up. 'Who's this for?' I was looking out for the movie star or politician to appear. "Your car, sir." Trying to look as if a chauffeur-driven limo picked me up all the time, I climbed in. "Where to, sir?" "Anywhere, James." Damn, the bloody windows were tinted and I wanted everybody to see me. We went up onto the hill tops surrounding Seoul to the Buddhist temples. The monks were not impressed with this bourgeois twit climbing out of the limo with flag and tinted windows. I can't understand some movie stars, they start off wanting all the glare and publicity that they can get, then they don't want to be seen or sign any autographs. I had a bag of Bic Clic pens and loads of notepaper at the ready just in case anybody wanted mine. One of the monks mouthed something. I could have sworn it was: "Him, big plick with Bic Clics!" I lit a few candles and nodded to some more bored monks.

At one temple, I bumped into two Europeans who happened to be female Swedish nurses and looked like the girls out of ABBA. They told me that they were working for an adoption organisation in Stockholm and had come to bring back two Korean orphans. "You guys want a lift? I can drop you off, it's no trouble," I said as I spotted the limo and chauffeur with the flag fluttering in the breeze. I was just about to casually point it out to them, when I stumbled and fell on my arse! 'Christ, you stupid Bic-plic.' I didn't know that monks were allowed to laugh at people's misfortunes but they did. The driver must have sensed I was trying to make an impression with the Swedes. He drove up, or I should say, 'glided the car towards us,' got out and opened the doors. Was this cool or what? On the way to the airport the next day, I looked out of the car window at the paddy fields and the poor sods up to their knees in muddy water.

'Some place, now for Libya!'

Me and my trusty limo driver.

LIBYA, 1ST TRIP 1982

(In which I survive Desert Brew and find treasures in the sand.)

It was several years later when I returned to Libya, working for the main Libyan oil company. So the conditions, I assumed, should be much better than what I'd experienced before. I should have remembered the old adage: "To assume, makes an Ass of U & Me." Read on.

The week before I flew out, Maggie Thatcher had given the Americans permission for their bombers to take off from the UK, to bomb Libya. 'Great.' I called into the local to see my mates before I flew out. They, of course, gave me their thoughts and opinions by telling me that "I must be off my head going there right after the bombing, and with the UK being an accomplice in the deadly proxy game." Or words similar to that effect.

Being off my head, I flew out again, remembering this time to put my bags on the plane myself. I was feeling apprehensive, as I didn't quite know what to expect. I walked through Tripoli arrivals and passport control. On display around the walls were photographs of dead or badly injured Libyans, women and children. Gaddafi's henchmen were waiting menacingly. I had calculated that going to Libya after the bombings should be almost OK, because the Libyans rely on ex-pats to run and maintain their oil fields. Basically I was a little concerned, aka shitting myself, but I thought, 'They know that it wasn't me personally that did the bombings... don't they?' Nevertheless, it was a stressful experience. I was stopped by a fifteen-year-old pretending to be a soldier, who snatched my passport out of my hand and studied it with a scowl. I was just about to say, "It's upside down," when one of the real soldiers came over. 'Oh shit,' I thought but to my amazement and delight, he snatched the passport off the brat, clouted him and looked inside. He noticed that I had been before, handed it back to me and said, "Have a good trip." Well, I guessed that he said that, at least I hoped.

I was met by the ubiquitous company driver, who took me to the ubiquitous guest house. He said, "I come back bukra, tomorrow, 06:30 for airport." It was a huge house, which seemed even bigger with me being the only one there. I put my bags down and immediately felt the dreadful feelings of homesickness. It has happened to me many times before. I've been at Heathrow, Gatwick or Manchester airports waiting for a long-haul flight, when this would hit me and I just wanted to turn around and go back home but couldn't, as financial circumstances dictated. I can imagine what it must have been like in the great wars when servicemen and women left their families to go to war. They didn't know if, or when, they would ever see them again. This wasn't quite the same of course but I could sympathise. At such times, I'd head for the airport bar to get some Dutch courage. This usually did the trick but the only problem was, when it wore off and you were in Libya, a dry country, you couldn't just go to the pub for a hangover cure and find some solace.

In the guest house, there was nothing in the refrigerator and I was famished. No TV or radio. 'God, this is going to be a long night,' I thought. I sat reading one of the newspapers that I'd brought with me, which was full of bad news, which didn't help my situation, then went outside onto the flat roof to get some air and look around. Suddenly, I heard somebody call me from the adjacent house. He looked European but I couldn't quite make out what he was saying, until he made the universal hand to mouth movement of 'Fancy a drink?' I was over like a shot. He was an Austrian engineer, working for Siemens, the huge German engineering company. He and his wife had been in Libya for two years and were leaving next month. His house boasted a huge home-made bar and a very healthy stock of beer and spirits, all home-made by his wife who was an expert. I suppose it helps coming from Austria! After a few beers, followed by a few of his funny tales, my homesickness wore off, at least for that night. I thanked the couple for saving me, as I am sure that I would have gone crazy in that house.

The next day, the driver arrived on time to take me to the internal airport to catch the desert flight. The usual mixture of Indians, Pakistanis, Filipinos, Brits, Americans, Yemenis, Thingummies, were all there. I recognised a face in the crowd: it was an Aussie that I'd worked with before. This bloke, Alan Holman, is in the Guinness World Records book for kayaking down the Amazon.
"You did what?" I'd asked.
"Well, it was only a little boat trip, sport."
He, of course, had been in the Australian Special Forces, the equivalent of UK's SAS and America's SEALs. Imagine the crocodiles, anacondas and all the nasties that can bite you on that journey!
"I met that old French bloke and his Sheila on my way and was invited for some nosh and a few tinnies, on their boat, the eh, can't think of its name?"
"Calypso," I chipped in, eyes widening.
"Yeah, that's right, mate, the Calypso!"
"Blimey, mate!" (I was into the Australianisms now.) "Jacques Cousteau, the legendary diver, explorer of the deep!"

"Too right it was, mate, the one and only. What a stroke of luck, hey? I almost ran out of supplies, didn't I? Old Cousteau pulled me out of the shit, filled me boat. Nice guy. I tried that Pernod shit, taste like cat piss."

"So you enjoyed yourself, then?"

"Too right, mate!"

(Alan gave me a signed copy of his book, 'White River, Brown Water', at the airport when we were both heading out to the desert for our second tour, a book which I still have to this day and recommend if you can get your hands on it.)

It was a good job I had our conversation to distract me as the turbulence and the engine-stall warning alarm, sounding constantly throughout the flight, didn't fill me with joy. We were about halfway to our destination when we landed to let some locals and oil-drilling roustabouts off. We all had to get off the plane and wait in the heat without anything to drink and no shade. I got talking to a Brit who worked for Rolls-Royce aero-engines.

"Was that the engine-stall warning alarm that was sounding back there?" I asked.

"It sure was," came the reply. "These bastards train their pilots on passenger flights."

Of course pilots have to fly for the first time with passengers but a bunch of raw rookies playing around with stall speeds, while they have a plane load of nervous passengers to consider?! They probably did this on purpose, with us being Western infidels. I was not looking forward to the remainder of the flight.

Sitting next to me was a young instruments engineer from the UK, who had emigrated to Canada to make his fortune. So the land of milk and honey didn't quite live up to its reputation and expectations! There were several of these people out in Libya, I was later to find out. Fancy emigrating to Canada and having to work in Libya. Rather defeats the object. On our descent,

I could see my home for the next six weeks. I said to the guy next to me whose name escapes me: "Be careful who you trust here, or better still, don't trust anybody." Having been to a place like this before, there are the usual cliques and staff personnel who think that they are a cut above the rest and usually never let the contractors into their fold. How right I was to be.

The accommodation was a mess. It reeked of cigarettes, the table was full of empty bottles, apart from the gunge left over from the home-brew kits. 'Jesus, who the hell lives here?' kind of went through my mind. The ashtrays looked as if they had scooped up the debris from a nuclear explosion. Lunch was nothing to write home about, to be honest it was diabolical; only the locals ate there. The others, I was told, brought most of their food with them. The dish of the day was, I think, camel. I had six weeks of this before I could start to bring my own food. Well, I thought, if Wilfred Thesiger the explorer could do it, so could I.

In the afternoon I met some of my new colleagues and was shown around the camp and workshops. After work, I headed back to my room. As I opened the door, I was enveloped in a cloud of smoke. My new roommate, Eric Wardrobe (my nickname for him, you'll find out why soon enough) was from Hartlepool – a monkey hanger, another nickname. (Monkey-hanger, if you don't know, is a legend I've heard many times: during the Napoleonic war, a French ship ran aground off Hartlepool and a monkey was the only survivor. Never having seen the French before, the locals thought the monkey was a French spy and hanged it. So as you can imagine, they don't really like the nickname but it's only a laugh and you take it on the chin. Coming from Liverpool, we have our own crosses to bear as it's universally assumed all Scousers are thieves.) Anyway, Eric Wardrobe was already into his forth cigarette and by the look of the empty bottles, the same amount of beer too. "Help yourself to a beer."

"Cheers," I said and accepted a very welcome cold beer. After a few more cigarettes and beers, Eric began to cook a meal in-between swirling his false teeth around his mouth, drinking, smoking and coughing up what ever he could from his boots.

"Fancy, cough, cough, wheeze, a bite to cough, cough, to eat?"

"Er, no thanks, not very hungry just at the moment," I lied. I was starving. I watched Eric eat/chew/suck his steak and chips minus his teeth. He must have serrated gums, I thought. I made an excuse and slinked out to Abdul's greasy spoon. On my return, I found Eric, mouth open, snoring his head off. I noticed that a cigarette had fallen onto his bed sheet and had burned a hole in it. The room stank of stale beer, cigarette smoke and grease. During the night: 'Bang, smash.' What the...? I awoke to find him standing in front of a now-broken wardrobe mirror. He had got up to use the bathroom and had opened the wardrobe door instead of the bathroom door.

"Some bastard tried to stick one on me!" he said in his Hartlepudlian gummy twang. He had thought his reflection in his wardrobe mirror was an intruder, so he punched it. Tomorrow, I thought, I would be moving out. Eric was a nice person, in a funny peculiar sort of way, but who could put up with that? Besides, I didn't want to die earlier than required, due to his smoking or in a bed-sheet fire!

The next day started with the supervisor giving each one of us a job to do. I was to spend the first week going out each day with a different person to show me the ropes and of course give me directions, or at least, tell me how not to get lost - easily done in the desert! The desert is a place that you either like or hate. It began to grow on me. I used to stop the truck miles from anywhere and listen. Hardly a sound, only a whisper of a slight breeze with nobody about. In some areas of the desert, when you walked it sounded hollow, like walking on a huge empty oil tank. I suppose you were, in a sense, as we were extracting oil from the ground. I remember passing a strange area where there were a number of petrified trees. Trees that had, over millions of years, turned into stone by the process of petrification as parts of the Sahara had been under water which would have brought sediments and excluded oxygen, thus preventing the wood from breathing and decaying. Although it is fragile, it feels like steel. I started collecting all sorts of things: petrified wood (some with knots in), many fossils, sharks' teeth, desert roses, arrowheads, shells and some unusual stones, which were smooth and had holes in the centre, where the sand over countless years had swirled and worn a hole in them. I still have this collection at home. I dug up an area about a metre square and found the sand to be many colours: red, green, white, brown and yellow - incredible. Soon, I was getting to like the place and the routine. One day, I stopped the truck to get some water that was in the rear. I was only stationary for seconds when overhead came several circling vultures. Waiting for dinner, I gathered. It was a sight you see in movies. Needless to say, I didn't stop too long!

There were tennis courts in our camp and after work I would have a game or two. One guy who used to play must have been around seventy but was very fit, a Londoner who had only known army life and, of course, the desert. He had spent most of his life institutionalised – what a life. After retiring from the army, he had found himself a nice little number in Libya with the Libyan oil company. To this day, I could never fathom out what his job was, as he used to do all sorts of things. Army life had suited him and, of course, the next best thing was working in the desert, where he was told what to do and given his accommodation, laundry and food. I would imagine he'd be like a fish out of water and completely disoriented outside of this type of environment.

There was an Austrian, who was one of the laziest, useless persons that I have ever come across and that's saying something having worked in many countries. He wouldn't lift a finger. His room was bare and depressing; he didn't have any posters or pictures on the walls, no photos or a calendar, and he had been there for years on a 28/28 rota. His room was like a prison cell. Fat Harry was another character – he was fat! And loved food. He would spend most of the day planning his evening meal. Poor Harry popped his frying pan not long after my arrival and went to the great larder in the sky. So I didn't really get to know him. They held an auction on his behalf. Harry had been a supreme hoarder of many things, especially cooking utensils. The bidding started and I stupidly started bidding for a glass beer mug and ended up forking out 70 quid for it. Golden rule: Don't go to an auction with drink and pity in your head. At least the money went to his family, so the pain wasn't so bad. A lot of money was made and lost that night.

Imagine being called 'The Rat'! The guy revelled in that name and he was a rat too, a Geordie rat, another who had unsuccessfully emigrated to Canada to a wonderful new life. The Rat was one of the most obnoxious, sly, back-stabbing, smug little shits I have ever come across and that is saying something, as there has been quite a few in my life. (Bullshit Bennet, for instance - you'll meet this character in Russia.) When the supervisor went on a field break, one of these staff characters would take over. We would end up having a puffed-up, smug bastard in what they called 'The Chair', full of self-importance at being in charge of half a dozen men. I was reaching forty and the Rat used to bring up the subject of heart attacks into the conversation every time he saw me. He reckoned that he had passed the dangerous age and I was approaching it, therefore I'd better watch myself. How Ratty worked that out, he wouldn't say. I think he believed that one only had heart attacks at a specific age and as he was now past this age, he was now safe, as he popped another cigarette is his mouth.

Sometimes, I would venture to the airstrip hanger and have a chat with the aeroplane mechanic, whose hobby was building and flying model aeroplanes. One day he was out flying his model plane, when the real pilot came over to watch. After watching for a while, he came over with his now-itchy fingers. "Let's have a go," he asked the mechanic. The pilot, a Canadian in his early thirties, was all brash and swagger. For his party trick, when he landed the real plane, which was a turbo prop, he would head straight into the hanger without stopping on the runway and park up for the night. After a brief instruction from mechanic to pilot on how to fly the model plane, it was soon airborne. I couldn't believe what happened next. It flew straight up, banked and nosedived into the ground. A look of sheer horror came to both of their faces. I was stunned too and embarrassed for the real pilot. The mechanic must have spent many long hours building it. Now it was completely destroyed and beyond repair.

The main hobby for most of the people, however, was not plane-making but beer-making. Some treated it as an art. They would look at it, smell it, swish it around their mouth as you do with wine in a restaurant. Some brews were stronger than others – you would have to be careful when going back to your room after a drinking desert brew. If you fell over, knocked yourself out or fell asleep in the sand, one of the desert nasties could get you, be it a scorpion, sidewinder, viper or a camel spider. Camel spiders can grow to the size of your hand and legend has it that they can jump several feet into the air and traverse desert sands at about 5mph. Now the nice bit: they eat the stomachs of camels from the inside out, hence the name camel spider. Although they are supposed to be harmless to humans, a bite can be very painful and requires medical attention before it can turn septic. They also have a powerful anaesthetic that numbs their victims (thus allowing them to gnaw at living, immobilised animals or drunks without being noticed). I do suppose one would if one was being eaten alive, wouldn't one? So one tried not to fall asleep in the sand otherwise it would be a hell of a hangover cure. Talk of hair of the dog that bit you; this would be the 'spider that bit and ate you'. What charming and lovable creatures they are? Legend, rumours or myth, I did my best not to turn it into reality! Imagine waking up minus a face!

The first thing one had to do before becoming a master desert brewer was to set up shop. It took many sorties into the desert to obtain the required bottles. The main source were abandoned camps where you could get empty litre-size milk bottles. The next step into the brewery trade was to thoroughly clean them. If they were not spotless, the beer would soon go off and so would you! You would then have to get something to brew your beer in. We used big metal water containers with a tap on the bottom. Then you would need a capper and caps. Last of all, the easy part - bringing the home-brew kits into Libya, which seems to be a contradiction/paradox in itself but was nevertheless true. Apparently, the trick was to visit a pharmacy in Tripoli or Benghazi, if it was open, and buy a case of bio-malt (two dozen cans, if you could). At the internal airport, the customs would search you, and out would pop the case of bio malt. "What's this?" they would ask. You would pat your tummy, pull a face and say, "Problem, Mr." They would look at you and probably say in Arabic: "You lying bastard." Well, those were the days. But if you were caught selling or giving it to the locals... well, that's a different story. A big problem was: one needed sugar and how could one get the sugar from the Libyan cooks without giving them a bottle or two? "You gotta give a bottle or two, you gotta give a bottle or two." Ron Moody, 'Oliver', Lionel Bart. Sorry about that. They knew what it was for so you couldn't hide the fact. So long as you didn't overdo it and get them pissed. What many people used to do was to bring back goodies, cheeses and all sorts of nibbles, and invite friends round to sample their fare. It became rather competitive, sampling each other's brew. The cheapskates were always found out. They were the ones that didn't bring the 'Pringles or good cheeses'. It was good fun and another way to while away the hours in the desert.

Peter, a guy from Manchester, invited me over for a drink and to listen to the FA Cup final between Everton and Liverpool. Everton scored, I jumped up and clasped his head between my two hands rather overexcitedly. Peter was deaf and was wearing glasses, the type that had hearing aids attached – he was not pleased! I had moved to another room and had to be diplomatic with Eric, my now ex-roommate as to why I moved out. I told him that he was a "drunken, chain-smoking, smelly, untidy, gummy fire hazard." Well, not quite. I told him that I couldn't stand the 'smoke'. That did the trick. Talk about going from one extreme to the other. My new roommate reminded me of the late Leonard Rossiter, who played the character Rigsby in TV's 'Rising Damp.' He was one of those Brits who had emigrated to Canada and now spoke with a bad Canadian accent which would grate on you. He was so tidy and fastidious he used to drive me up the wall. I had been smoked out and now I was being cleaned out. He once told me that when he and his wife invited their friends to dinner, him and his mate - no kidding - used to dress up as waiters - white shirts, bow ties and black trousers - and serve the drinks and dinner to their wives. What an arse! To this day, I will try my utmost not to share a room again.

After six long weeks, I sat on board the British Caledonian plane or 'B-Cal' as they were called back in those days. Gaddafi's henchmen would come down the aisles looking at everybody as if we were bank robbers. Why they did this I don't know, probably just to wind people up. I was well and truly ready to fly out and as soon as the wheels left the ground there would be a huge cheer and the air hostesses would get the drinks out! Ask Kate Adie the reporter, if you ever see her, what Libya and the people are like, in the towns not the desert. As in most places around the globe, the ordinary man in the street is usually OK. It's the rulers and politicians that are the baddies! On reflection, apart from the arseholes and weirdoes that I met on life's fly-paths in Libya, I found working in the Sahara was a great experience that I will treasure

Baijy, Iraq, 2nd Trip 1983

(The Darlington Map Reader and the mini World Cup.)

I drove to Darlington for an interview at the office of William Press International for a project in Northern Iraq. I couldn't help but notice a large map of Iraq spread out on the table. The interviewers entered the room, introduced themselves and sat down. All seemed to be going well, until one of them said: "Cast your eyes on the map in front of you." He pointed with a ruler. "This is where the hostilities between Iran and Iraq are taking place, waaay down south. And waaay up north is where the refinery is located." He immediately got up my nose with this statement. I reached over and pointed out that the Iran/Iraq border runs the full length of Iraq and that it wouldn't take a jet very long to shoot across going from east to west.

This retort took them by surprise, so I carried on. "What type of equipment do the Iraqis have on their plant?" I was being a smart-arse but they were trying to outmanoeuvre me. It is a war zone and they were trying to make it sound like Disneyland. "Er, you will find out when you get there." They obviously didn't have a clue. This was one of those non-technical interviews given by bean-counters interested only in getting bodies out on their project and I was trying to conserve mine, having previously been to Iraq and experienced a narrow escape with my life. "What salary had you in mind?"

I gave them a ball-park figure, taking into account that I would be returning to a war zone. "I don't know if we can pay that much." I never answered but left it to them. "It's going to stretch us… but OK, seeing you have the background and experience that is required and you have been to Iraq before, when can you go?" A week later, I arrived at Saddam Hussein International Airport, Baghdad.

After I had checked into the hotel, I went down for dinner, to meet up with my new colleagues from the project. Not to my surprise, they were still in the bar with one of the devious map-reading accountants who had interviewed me in Darlington. The following day, we travelled north by coach, which took us the best part of three hours. As we passed the refinery, I noticed gun-emplacements dotted around the perimeter wall. 'Oh the war is way down south is it?' I thought. As the camp was only about a half a mile away from the plant, I wondered why they hadn't economised even more and had us live on site. The accommodation was like a prisoner-of-war camp, no better than ex-army billets; the rooms were featureless and being next to each other was going to be a problem, as I was later to find out. With impeccable timing, nothing was ready for us. "Er, where do we eat?" "I think they may have something laid on at the refinery?" We ended up eating cold buffets for the next week in the refinery restaurant until our restaurant was ready.

On our first day at work, we were told to write out our CVs, which I thought was very odd as I had never been asked to do that before on site. Trying to remember jobs, dates, equipment and locations was a real pain as I have worked on many projects in many countries. However, I did it and took an instant disliking to the person doing all the talking, or in his case, the 'mouthing'. He was a Palestinian and had an attitude problem. To be honest, he was a very aggressive, bad-tempered, arrogant, ignoramus little sod! The job also turned out to be odd as we didn't have much to do. I wondered why they wanted us here. They had their own qualified engineers who were most capable of doing the work. Were we some kind of insurance? Did they think that the Iranians wouldn't bomb the refinery because a small team of British engineers were now working there?! We all felt a wee bit superfluous, to say the least.

The camp bar opened up. Turkish beer only, probably full of formaldehyde. As there were some supercilious old farts amongst the British contingent, it would at least serve its purpose and preserve them. Our Polish technicians finally arrived and could hardly speak a word of English and of course none of us could speak Polish. How were we supposed to communicate? This project had really been put together well. The Poles would all congregate in the lounge part, while we stood at the bar, but oddly none of them had a drink. I asked the Polish barman, who could speak some English, why they were not drinking. Of course, they didn't have any money. How could we possibly drink in front of them? And if we did offer them a drink, how embarrassing would that be? There was no way that any of us could drink in front of them and we decided to send them drinks in any case. You would have thought they'd won the lottery. A few came over and joined us. They turned out to be a very friendly bunch, even though we barely understood each other. The English-speaking barman was kept rather busy that night.

After a few drinks, a few chinks in people's armour soon surfaced. One in particular was Alan, a real Jekyll & Hyde character from Manchester. One of the things that we had to do at the refinery was to recommend spare parts and the percentages required, taking into account that they were still at war with Iran. I found myself sharing a desk with him. We were trying to estimate how many spare boiler tubes would be required in an emergency – and emergency could mean bomb damage. Bloody hell. The boilers were the size of an average detached house and were the heart of the refinery. "I would say at least 20 percent extra, what do you think, Alan?" No reply. "Alan." I looked at him. He had this strange look on his face; he already looked strange without having an even stranger look on his face! "Alan, you OK?" He started to shake and foam was now coming out of his mouth. Oh fuck. He was having a fit. Next thing, a huge Iraqi picked him up and slung him onto his shoulder like a sack of potatoes. Alan not only looked like a garden gnome, he was not much bigger than one. I followed Alan's arse as it was bundled down several flights of stairs. Where is this arse taking the other arse? He laid him (dropped him) on the ground outside. Ouch! Alan was the colour of boiled vomit, still foaming merrily away – and there was absolutely no way I was giving the kiss of life to a foaming garden gnome – when all of a sudden he stood up, burped and said "50 percent", as if nothing had happened.

Kishorn is a remote group of Scottish villages on the northwest coast of Scotland, a very beautiful and tranquil place. The company, Howard Doris, have a shipyard there, where they repair and upgrade oil rigs and drink whisky and eat children for breakfast etc. BP had two bars at their base there. One was the Staff Bar, for collar-and-tie and plummy-voice types and the other was the Bears Bar, for painters, scaffolders, riggers, labourers, laggers - a blood-sweat-and-sawdust-on-the-floor type of establishment for discussing art and religion and where swearing was mandatory. I later found out that Alan had one night been gnoming in the gloaming around Kishorn and, whilst pissed, had ventured into the Bears Bar and opened the valve on a firehose, dousing the chaps inside, as you do, and bringing their discussion on the meaning of life and Tolstoy abruptly to an end! I won't go into any further detail about his fire-fighting escapade but will leave it to your imagination, as a crowd of wet, drunken North Sea Tigers asked him kindly, through their now doused, soggy, cigarette-holding mouths, "Would he be awfully kind and mind turning it orf?" They then discussed with him the pitfalls of what he had just done. And there's more. Our intrepid gnome was gnoming again on board a small oil-company plane. He attempted to go for a pee but instead of opening the toilet door, tried to open the exit door of the plane, much to the annoyance of fellow passengers. He was a nice person in a funny sort of way really, until he inhaled the smell of a barman's apron. He got pissed just ordering a drink.

I had the unfortunate experience of being at the camp bar one night, when he walked in and, in gentlemanly tones, said, "Gimme a fucking beer," to the poor Polish barman. 'Oh dear, he's been at the aperitifs again.' What one had to do was surreptitiously move all the bottles and glasses out of his reach and cover one's groin area with one's leg and hope and pray that he would fuck off! This night somebody had foolishly left a beer bottle within his grasp. Before anybody could stop him, he hurled it across the room where the Poles were sitting. By a pure miracle, it didn't hit anybody but landed on one of the glass-topped tables. As glass-topped tables have a tendency to shatter when hit by a bottle thrown by a drunken, bad-tempered, pissed gnome on a Saturday night, all hell let loose. "Christ, get that silly c*nt out of here before he wrecks the place," I thought I heard somebody exclaim. It might have been me.

Two of us grabbed hold of him and gnome-handled him back to his room. He didn't take kindly to this and took a swing at my fellow gnome-handler, knocking his glasses clean off. Now this chap was a mild-mannered PhD person who became ill-mannered when gnomes punched him in the chops. He thumped Alan, who obligingly fell asleep, giving us just enough time to get him back to his room and throw him onto his bed. No sooner had he landed on the bed when he sprung back up, as if he was on a spring. 'Boing!' Just like the Hollywood actress Glenn Close in Fatal Attraction, in the scene were she is supposedly strangled and drowned in the bath by Michael Douglas and then sits up with those big staring eyes. Before we could get out of the room, the little bastard started hurling cactus plants at us, including the pots. The Doc put his arm up to stop one hitting him in the face and ended up with an armful of needles that took ages to extract. The first week that I was in Iraq, I was just dozing off to sleep when I heard wailing, just like a banshee. 'What the hell was that?' It was the garden gnome, having a gnome-mare.

Another character was a tall, fat Londoner who used to walk into the bar, shirt collar turned up, hair slicked back Teddy-boy style, and his name was Dave. But of course, what else? He was a cockney wide-boy, basically a bullshitting bastard, and had the most irritating way of expressing himself. If he thought that he'd said something smart or cool, he would cover his top lip with his tongue, whilst moving his head up and down, like one of those nodding dogs in the rear window of the car in front of you. Looking at everybody, he'd then wait for applause or someone to say: "Gee, Dave, that's cool, how do you know these things, isn't he amazing?" It didn't take the Iraqis long to suss him out. They gave him the nickname, aka 'Cloutchy'. Iraqi for wide-boy. One day in the office, he said to me, "Hey, Scouse, mind if I ask you a question?" "No." He asked me what my salary was and, like a fool, I told him. Big mistake. He stood up, flung his car keys on the table and stormed out. I had obviously touched a raw nerve. Later, I was to find out that he tried, on several occasions, to get me fired off the project. One of my worst traits in life is I trust people too much, like the cactus-throwing gnome. Dave could be as nice as pie and witty too, but drop your guard at your peril. As soon as you turned your back, the knife would be out. And how did I know this? His own colleagues and the Iraqis used to bloody tell me!

There were also two Australian guys who were running the Japanese camp and cleaning up financially as well. We used to call around to their place for a drink and a chat. It was just like the movie *A Town Like Alice*. They were really smashing guys who worked their butts off and screwed the Japanese corporation with whom they had a contract. One of them was born in Coventry, UK but had emigrated to Australia when he was a nipper. I used to pull his leg by asking: "How could you possibly be Australian when you were born in the UK?" This really used to piss him off. I know that you can change your allegiance to another country and take their nationality but it still doesn't seem right to me, but there you are. He brought me two Aussie hats and a hunting boomerang when he came back from his leave. So he didn't take it to heart.

Some weekends, I would walk up into the surrounding hills where I'd come across mountain people who were extremely poor and living rough. They would ask me for medicines etc! I started taking them aspirins and ointment, antiseptic creams and dressings that I had scrounged from our medic and things that go to waste in your bathroom. You would have thought you'd given them gold. Southern Iraq had its marshes and its uniqueness; what it is like now, since Saddam Hussein drained the area, I don't know. Northern Iraq was completely different, very Turkish and in winter very cold with snow. I used to like going around the market places with all its spicy smells and bartering with some of the stallholders. A really wonderful and friendly race of people, who would later be gassed by Saddam Hussein and his henchmen. My, how the mighty fall, which he did in the end, through the trapdoor of the gallows.

Saddam came from a town called Tikrit, which was about halfway from Baghdad and our camp. I used to pass it a lot and one day I decided to have a look around there. You could sense the danger and the evil in the town. Most of Saddam's clan lived there and some of them used to visit our bar. One in particular carried a gun, which was bad enough but when he got pissed, he would take it out, point it at you and pretend to shoot you. And if he did, he would get a pat on the back from Saddam, probably a medal too, no questions asked. One night, this plonker pulled out his gun and I noticed the twinkle in Alan the gnome's eye. I bolted to my room smartish. As you can gather, one was inclined to get a wee bit jumpy with morons pointing loaded guns and drunken gnomes throwing bottles.

And then there was Gulliver. Gulliver was this guy's real name. He reminded me of Desperate Dan, the cow-pie eating cowboy out of 'The Dandy' – he certainly had his chin. Gulliver was about six foot eight and built like an outside toilet. He was the tallest and oldest child I knew. He would come into our office and all of a sudden it would be like an escape committee meeting at Colditz Castle. He had been making escape plans in case the Iranians invaded. "We could store fuel, water, tins of food and head for the Jordanian border." Your imagination would start to run away with you as you looked out of the window onto the mountains, visualising Iranian tanks and soldiers coming over the top. He would go on about this all the time and, after being involved at the start of this war, it was beginning to wear thin with me.

The air raids did no good for one's nerves either. There would be at least two per week, as the Iranians were trying to bomb the refinery. I thought about that lying map-reading bastard in Darlington... I was driving around the refinery one day with Dave the cockney spiv, when we came under attack. The air-raid sirens started up, people were running everywhere. "Scouse, you've been in this situation before, what do we do?" Without answering, I changed down gear, booted the pedal and took off, my thinking being the safest place to be was definitely not on an oil refinery! I drove to the main gate on two wheels and, to my horror, found it shut. The Iraqis had locked us in. We were sitting on a bomb, waiting for a bomb...

Not only did we have the threat of being blown to kingdom come, but of Gulliver too! He had been on leave and had returned with this huge metal plate, as you do. It was two foot square. Instead of reading books, writing, listening to music or making model planes as normal people might do, he would be hammering and drilling holes in this plate, making trains or something, probably a shield. I used to dread sharing a vehicle with him because in the mornings, at the refinery gates, we would all be searched by the security guards/soldiers – well, they were at war, after all, and this was a strategic target, so it was only to be expected. As many people worked at the refinery, this would take some time. We were in the queue. Gulliver was getting twitchy and started to badmouth the Iraqi soldiers, who of course just happened to be under pressure and armed to the teeth with Kalashnikov AK-47 assault rifles, handguns and grenades attached to their belts. One of the soldiers heard his verbal abuse. Everybody in the world knows what 'fuck' means and, followed by 'rag-headed bastards', this did not go down too well at this particular moment. The soldier approached. Oh shit. "Shut up, Gulliver! Can't you see he's armed and angry with a licence to kill stupid c*nts like you, with the full blessing of Saddam?" He started arguing with the soldier and more of them approached, taking the safety catches of their guns. "Gulliver, can't you see that we are just about to die if you don't shut your mouth?" He soon got the point when the business end of an AK-47 was stuck up his nose! "Now, why did you do that? We are still getting paid, and in their time, we are just going to our office to make more escape plans." I was trying to cool him down before Gulliver went on his travels - to the pearly gates. The crazy situation was, about an hour after all the searchers, they would put a Dervish on the gate, an old and poor Muslim man who didn't know what day it was.

The following day, we all had to go for a cholera booster jab. We were sat in the waiting room when this young laboratory technician came past with a tray full of glass phials, filled with God knows what. They were cloudy and had bits floating in them. He grinned a toothless grin and made a gesture that we were about to be injected with the contents of these phials. There was no way that was going inside my body. An Iraqi doctor came out and gave him a Benny Hill slap, followed by a rollocking. We found out what was in the phials. They were urine samples. Perfect. Some years ago in Bahrain I had to give blood and urine samples to obtain a visa. They just used to stick a pin into one person, wipe it with a piece of cotton wool and stick it in the next person. Russian roulette had nothing on this. Imagine that today! You cannot get into many countries, even Africa, without being tested for HIV and other nasties. So you could end up having two full medicals per year. All the urine samples that were finished with just got tossed into a cardboard box and of course cardboard boxes cannot contain urine and leaked everywhere. On a visit to a hospital in Baghdad for an X-ray, I peeped into one of the wards – the stench was terrible. Many people in the UK complain about our NHS. They obviously haven't seen some other countries' medical facilities. The UK's health system is one of the best in the world, even with all its problems. Pity so many people abuse it!

One day I was watching some of the Poles playing five-a-side football in the camp car park and arranged a game with them. Of course we got hammered but it sowed the seed and was good fun. Not far from where we were was a much larger camp, which had a full-size football field. The camp housed personnel from countries such as Italy, Czechoslovakia, Poland, Egypt, Yugoslavia, Germany, India and Iraq. My intentions were to arrange some friendly games between us. I approached the management from each company, who liked the idea and offered to sponsor the games by donating money to buy football kits. As we only had a handful of Brits of playing age and a similar amount of Americans in our camp, we had to combine and play as UK in one game and the USA the next. At the opening game, they gave us the honour of playing first. We arrived at the football ground all togged out in our new kit and couldn't believe what greeted us. There were literally hundreds of spectators waiting in anticipation for the game to start.

It was 7:30pm and the floodlights were already switched on. The referee and his two linesmen were kitted out like in a professional game. The two teams stood in a line, just as they do at Wembley, and they started playing our national anthem. This kickaround was suddenly becoming a mini World Cup; the only things missing were Alan Hansen, Alan Shearer, that crisp guy with the big ears and the Jules Rimet cup. None of us were fit and we got stuffed by the Czechs but they were younger and were used to playing on their own ground. Well, that was our excuse. When we didn't have a game to play, we went along to watch. One game in particular was the Iraq v Egypt match. A goal was disallowed for Egypt with only minutes to go, and some of the Egyptians stormed the pitch and a big punch-up started. This was the real stuff, our World Cup had well and truly started!

Our next game was USA v Italy. We turned up as Yanks and sure enough, both anthems were played, followed by a few cat-calls from the crowd. The game got on its way. We were only two goals down when I slipped going for the ball and fell awkwardly on my back. Ironically, I had my fist clenched and fell on it. I ended up being laid up in bed for two days. The mini World Cup was a huge success. It got people out of their rooms and talking to each other. It was good for international relations too. Which is why the real World Cup is such a good thing. The Eastern Bloc workers could see for themselves that we from the West weren't demons and vice versa. We received invitations to have a drink in their bars and we of course reciprocated. Apart from the language barrier, everybody enjoyed the football, the atmosphere, the beer and bonhomie. Czechoslovakia won the cup; we came last. And blamed the Yanks, who blamed the Brits! We never quite resolved that issue, but in the end who cared?

Polish interpreters arrived, two men and two women, which was a blessing. But it wasn't a cure, just a quick fix. We now had the situation where, if there was a problem at the refinery, an Iraqi engineer would come to you for assistance and convey the problem in broken English. You would then have to explain to an interpreter what you wanted the technicians to do. They would then translate and relay your instructions to the engineers and technicians. Simple? No! As the interpreters did not have a technical background, things could go very wrong, so you had to keep your eye on things most of the time. I remember asking a Polish technician to remove a valve from an isolated steam boiler (the boiler being about the same size as a detached house). Just by chance I was passing by when I saw him working on a live boiler that was under full pressure. It would have killed him for sure. What ever happened to Esperanto? Why haven't we, in the 21st century, got a world language? I know English is a prime language throughout the world but try telling that to the French. One interpreter with whom I got on well with was disliked by the other Poles, as they were wary of him. They suspected that he was a 'plant' on the plant, a spy from their secret police sent to spy on them (good old communism). Imagine not being allowed to leave your country without a minder and people spying and reporting you and living under a tyrannous regime. Not much money, not much hope.

However, relations were beginning to improve. The Poles were getting on well with their English, as they were now having classroom lessons and we were picking up useful Polish words. We could now have a laugh and a joke together without blowing the place up. One day, we were watching

Iraqi firemen training on a new and very expensive Swiss fire truck that had only just arrived. They connected the hoses and were waiting for the water to flow, standing flat-footed with feet parallel, instead of leaning forward with one leg in front of the other. We knew what was coming and it did. The water came through under pressure and sure enough they all went on their arses with the hoses snaking everywhere. On another occasion, one of the warehouses caught fire and they appeared in sparkling new fire tender with sirens blasting, lights flashing. They parked the tender right next to a fire hydrant, obstructing it and then couldn't connect the hoses. Then they couldn't start the fire tender's engine to move it out of the way of the hydrant. Of course the tender was empty too as they hadn't refilled it after their fire drill. Hence, the warehouse burned down and the fire tender got blisters.

The language barrier was a big problem, not only with the Poles but also with the Iraqis. I had given the Polish technicians the job of fitting new V-belts to some huge fin-fans, 12 foot from tip to tip. Some time later, I went to see how they were getting on, remembering what had happened in past projects. This time I'd gone through the permitting and isolating procedures with them. In this case, the fans would be wracked out, a procedure where a certified electrical engineer would remove the fans' electrical equipment in the sub-station so that they couldn't be started until the job was completed and signed off. Also, they would lock the starting equipment's cabinet key away and that would require a signature to get it out. Easy peasy? No! As I was standing there, I noticed an Iraqi electrician go into the sub-station. I had a gut feeling that somehow something was wrong. I told the technicians to come out of the fan cage. No sooner had they alighted the cage, when 'whoosh' the fans started! They would have been chips. What had happened was, the Iraqi electrician had gone into the sub-station, seen the equipment isolated and, without any thought or questioning as to why, had put it all back again. So another lesson was learned: keep the fuses on your person or locked away in your desk.

"Has anyone seen the analyser engineer?" He had arrived two days ago and had not come out of his room. He was suffering from the DTs, the dreaded 'delirium tremens' and he had it good, or I should say 'bad'. He was the most unlikely instruments analyser engineer that I have come across. He was built like a Vulcan bomber and was an ex-undisputed combined services heavyweight boxing champion and no doubt a Glasgow street fighter as well. Instrument engineers play with little screwdrivers and tweak clocks and gauges and things with delicate hands. His hands were like Christmas hams! What was King Kong doing here as a tweaker? He liked a wee dram or two too. How unusual. He later confessed that he was an alcoholic and that his brother had died from alcoholic poisoning. At work, or when he showed up, he always used to smell of a heavy cheap aftershave, obviously to disguise the smell of booze on his breath. He had those staring eyes, like a rabbit caught in a car headlights, and sweated profusely. A nice person and a very good engineer but booze was his master. I was having a beer and a chat with him one night, when he said, "Scouse, have you ever had a Rusty Nail?"

"I have trodden on a few."

"No, the drink."

"What's that?" Big mistake.

"Brandy and whisky mixed."

"Oh."

We headed back to his place and were soon into the RNs. As the night wore on, we started talking about boxing - another mistake. Ever been taught the rudiments of boxing by a drunken ex-heavyweight boxing champion? Thought not! We ended up so pissed that he started throwing punches at me, which stopped about one millimetre from my chin. If one of his Christmas hams had connected, my head would have been in orbit. I died a thousand deaths the next day and I have never touched a Rusty Nail again and never will. The drinking started to affect his work - on some days he wouldn't leave his room and would wait by his door to ask someone for some drink, or rather plead. He even used to plant drinks around the refinery to rid him of the DTs. The last I saw of him was when I drove him into Baghdad to the airport for his field break. He was in a bad way - agitated, sweaty – and shook all the way there. He almost ran to nearest bar at the airport. What a sad sight.

The project had for me come to an end and I was glad to be leaving as the Poles had become very morose. One couldn't blame them. After all they had been in Iraq a whole year without going home. Some of the backstabbing Brits who had sucked up and pleaded with the Iraqis to get an extension to their contracts stayed on. I witnessed my first moonrise on leaving Iraq. At least this time I left without the duress of my last venture in this violent part of the world. Some job, some place. Glad to go!

Christmas Day with my Polish Tech Team.

BENIN, WEST AFRICA
1985

*(Where I learn all about taxation and
how to handle a witch doctor.)*

"Who's the client?"

"Not really sure, but you'll find out when you get out there."

I was going to work in the heart of darkness and I did not know for whom - how odd. After giving my malaria prophylactics time to take effect, I flew out. During the flight to Geneva, en route to West Africa, I questioned my wisdom and decision to accept this bizarre contract. Was I doing the right thing? As the flight had been delayed at Heathrow, I was now running late for my connecting flight in Geneva. I informed one of the stewardesses of my predicament, hoping that she could get a message to Geneva. She returned and said, "When we land and before we stop, go to the front of the plane ready to disembark. A car will be waiting to take you straight to your connecting flight."

We landed. I unbuckled my seat belt as it was taxiing and went to the exit door. Another guy was doing the same thing. We both ran down the steps and climbed into the waiting car. Just as we were going to the plane, his phone rang.

"What's wrong?"

"I've got to go to our head office."

"Problems?"

"Yes, problems."

"No doubt, I will bump into you in some bar in Benin?"
"Yes, no doubt you will."
I found out later that he was one of the oil company's top men in Benin.

I was driven straight to the awaiting Air Afrique plane. I ran all the way along the departure tunnel, then climbed the steps of the awaiting plane and, before I got to my seat, we were on the roll. 'Phew.' A large G&T was my first request. I had to keep the mosquitoes away, calm down and take stock of the situation. I alighted from the plane, my first time in deep Africa. It was hot, sticky and smelly. After eventually collecting my luggage from the carousel, I headed straight to immigration. I usually get the vibes of a place at immigration. The official was about as friendly as a lion with a sore tooth who hadn't eaten for two weeks. A big, sweaty guy who, not too unlike the late and infamous Idi Amin, was searching through my baggage. Somehow I could tell that he wasn't using a deodorant. He asked what I had in the briefcase.
"Oh, just newspapers and private things."
"Let me see them." He took a bottle of my duty-free whisky.
"I see you like a tipple?" I cursed.
"This be mine. I take them."
"Why not? Just help yourself, as no doubt you will."

"Mister, mister, I porter your baggage?" Said by at least thirty hopefuls that had surrounded me. "No, go away." Where the hell was the bloody agent? Nobody was holding the usual bit of cardboard with name scribbled on it. I walked up and down the arrivals hall followed by the throng. "Mister, mister!"
"No, go away." Said in a not-too-polite way.
Alarm bells started to ring. "Shit." I was in Lomé, Togo, when I was supposed to be in Cotonou, Benin. How did the travel department mess this up? Nobody to meet me, God knows how far from my destination and I didn't even know

who the client was. I looked for a money changer but it was closed. I had about 100 in sterling and US$200. US$20 and sterling £10 notes were the smallest denomination that I carried. I went outside into the street; still no sign of the agent. About forty would-be porters lunged towards me: "Mister, mister, I carry?"

"Taxi, taxi!" I shouted. I started to feel a little bit agitated, to say the least. 'OK, get your act together, for Christ's sake.' I climbed into the nearest 'rent-a-wreck', followed by a cheeky streetwise urchin. "Where you go, mister?" I held up a US$20 bill, which was most probably two months pay to the taxi driver, and two years for the urchin. "Take me good hotel and no stop." Amazing what a bit of hysteria does to you. We drove down some very dodgy, darkened streets that had no street lighting. Not very London, not very anywhere. There were plenty of road-side vendors, selling cigarettes and, oddly enough, all kinds of classy spirits under the light of smoking oil lamps. Some of the crowd started to stare at me as we drove past, which didn't do the nerves much good. It took about half an hour to get to the hotel. At the check in, the hotel manager asked to be paid up front. I refused, suspecting I'd be presented with another bill in the morning too. "I no pay now." A bit of bravado was creeping in.

I showered, changed and went down to the bar for a beer and to try to get some information on how to get to Cotonou, Benin. I was immediately accosted by several prostitutes who seemed to come from out of the walls like the Clay Men from an old *Flash Gordon* movie. I quickly finished my beer, escaped to my room and locked the door. The next day, I went down for breakfast to assess my predicament. Benin was the next state to Togo, about thirty miles to the border. As I paid the bill to check out, I noticed a taxi tariff on the wall: hotel taxi. Great, at least I will only get robbed a bit. Even though there was a tariff, the driver still tried to acquire more money. After haggling, I agreed a tip, providing that he got me there in one piece. As this was my first visit to West Africa, I sat back and took in the sights, some good, mainly bad.

It had been raining and the roads were slushy and dangerous. I eventually arrived at the border and was ushered to what was no more than a mud hut.

"I be the chief of police of this place."

"Pleased to meet you, Chief." We shook hands, sat down looking at each other, wondering what the next move was. On a shelf, he had this huge ghetto blaster playing African music, at about 150 decibels. (As apposed to the acceptable 90 to 95 dB, the maximum for humans, but sustained exposure would result in hearing loss. 140 dB is a jet engine taking off! 194dB is the loudest sound possible.)

"Great music, Chief," I lied, as it reached 193 dB and my face starting cracking.

"You can have the tape."

He checked my passport yet again. Jesus, that bloody music. I wondered what was going to happen next. They had checked, ransacked my baggage and of course pilfered the remainder of my duty free. We sat there. I needed him to stamp my passport, to enable me to get through the border. If I offered him a bribe, knowing my luck, he maybe, just maybe, would be the one-in-a-thousand honest police chiefs in Africa. An idea came to me. I asked him if I had to pay any tax – this seemed to jolt him into action.

"Ah yes, the taxation, you have now to pay the tax."

I handed him a £10 note.

"What this? Only dollar give me."

It's amazing, in many countries they just do not recognise sterling and, as for the pound coins, forget it. They look at the notes as if the Queen was from Mars and had two heads. But the dollar soon brings a sparkle to their eyes - the bald eagle strikes again. The smallest US dollar note that I had was a $20. 'Shit.' I handed him one. The stamp was now in my passport before I could say 'taxation'.

I crossed no man's land, the bit between the borders, in order to get to the other border post/mud hut. The guards searched, ransacked my baggage, cursing as I had no more goodies for them to steal. As by now, all my duty free had departed through the departure gate. Yet again I was ushered, or rather pushed, into the police chief's office. A sudden feeling of déjà vu crept in, as my face began to crack due to the greeting of 140 decibels from this police chief's ghetto blaster. I think that they were in competition with each other.

"I be the chief of police of this place."
"Pleased to meet you, Chief." I checked my face again for further cracking.
"Sit down the chair. What you do here now? What your name?" We sat and stared. My ears screamed out for mercy at the pounding from the blaster.
"Do you need taxation to stamp my passport?"
His face lit up like a flare. "Ah yes, the taxation for the stamp."
I tried a £10 note. It was as if I had handed him a dog turd.
"What this? You give me dollar." It was the difference between a turd and a gold bar. I pocketed the turd, he pocketed the gold bar and stamped the passport. We shook hands. He gave me his card.
"You telephone me this number, I come to your hotel and take drink."
"Sure, Chief. Don't forget to bring your music."

After a very bumpy and slushy drive, I finally reached the hotel; in fact, it seemed the only decent hotel around, apart from a couple of shady joints that we passed. It was still early afternoon and the place was deserted. I took a stroll past the swimming pool and along the beach. The rain had cleared the air, so it was nice and cool. From my veranda, I could look down at the Atlantic rollers pounding the beach. This was like something out of a James Bond movie – it was gorgeous. After a nap and a shower, I was heading for the restaurant when a familiar face greeted me. I couldn't believe it. It was my old mate Jock, from Burntisland, Scotland, whom I'd worked with in Bahrain.

We downed a few beers and updated each other on where we had been. It had been some years since I had found him lying arse up, drunk as a skunk on the floor, in a pub car park in Bahrain. He had scared the living daylights out of me, as I'd thought he was dead or even worse for a Jock, mugged!
"Hey, we're all going into town tonight in about an hour's time, fancy coming?"
"Why not, ok. See you in the reception."
After the habitual haggle with the taxi driver - "We're not buying the bloody thing etc" – we headed into town, about twelve of us.
"Whose turn is it to be arrested tonight, then?"
"What do you mean?"
"Oh, nothing really," John said. "Just that these thieving arsehole soldiers are always on the take, wanting money, cigarettes, drink, anything they can get their grubby hands on. We take it in turn to be arrested; they just keep you by the side of the road and the others go back to the hotel for their passports and usually, by the time we get back, the arrested person is set free. We give em naff all, otherwise we'd get arrested everywhere we go, no peace."

John, who gave me this advice, was our security and safety manager. We arrived at the favoured bar and were immediately plagued by pimps and prostitutes and badgered by bloody street hawkers who came in their droves, carrying their wares. I was at the bar talking to a Taff, one of the electricians who had once been a mercenary in the Congo. His salutation to the hawkers was: "Fuck off, you black boyo c*nts." This time he was more courteous. He grabbed the ebony statue from the hawker and pushed him away. Then pulled out a huge

Crocodile Dundee knife from his belt and stuck it into the ebony statue. He began to drill a hole in the bottom of it with the knife. With ebony being very black, if any white showed, it would mean that the hawker had used ordinary wood and had dyed it, or had put black shoe polish on it. Lucky hawker never met his nemesis as no white appeared. Taff then grabbed an ivory statue, pulled out a cigarette lighter and proceeded to burn the bottom of the statue, to check that it wasn't plastic. It was the first of many African learning curves for a rookie.

Locals started pouring into the bar, wearing the oil company's baseball caps. I shall call the company Zippity Doo Dah Oil, for reasons that you will find out later. They were also carrying banners with the company logo and name on and, in addition, greeted us like their saviours and long-lost friends. What had happened? One of our accountants had found an old copy of the Financial Times in a drawer in their office. It read, sic, 'A Norwegian oil company, for the past five years had, under contract with the Benin government, developed and produced Benin's only oil property: the Benin Offshore Field. Life was rosy for their team of expatriates. Having spent over US$100 million on the first phase of the development, they had started to spend another US$50 million to boost production to 11,500 b/d (barrels per day) from 8,500 b/d. Then without any warning, Zippedy Doo Dah Oil announced a US$2.billion joint venture with the Benin government, part of which included investment in the oil field to increase production to 25,000 b/d. The Norwegians were taken completely by surprise and denied all knowledge of a change in operator, telling callers that they were confident that a mistake must have been made somewhere but there was no error.'

The Benin government confirmed the tie-up with Zippity DD in a half-page advertisement in leading financial newspapers around the world, leaving the Norwegians red-faced and ill-informed. Benin, formally the French colony of Dahomey, adjoins Nigeria, but has little hydrocarbon resources to compare with its neighbours. The government saw the deal with Zippy as a way to raise low living standards, because the US$2 billion joint-venture program would apparently cover most aspects of the economy, as well as taking over the oil field and more than tripling production. 'Oh I forgot, it would cover the President's pocket money too!' Zippy were going to undertake further offshore and onshore exploration for hard minerals as well as hydrocarbons. No wonder the locals were treating us like heroes. Zippy were also going to build refineries, hydro-electric and other power-generation facilities, fertiliser plants and irrigation systems. Oh and build and commission a new international airport. Some deal. No wonder the Vikings were somewhat pissed off, thinking of all the kroner that was still owed to them. To raise production to 25,000 b/d wasn't just a little pie in the sky, but completely unrealistic. What were they really doing here?

We headed back to the hotel and sure enough we were stopped. Here we go.
"You might as well take your turn now, Ken, it will stop you agonising as to what it'll be like."
"Sure, I am really agonising and can't wait. No problem." Dutch courage had crept in after a few beers. I climbed out of the taxi. The soldier pointed his machine gun at me. "Oh fuck." Where did that Dutch courage go when you needed it? There are only a few things in this life, not many, more scary than looking down the business end of a machine gun held by an illiterate, drunken or drugged African soldier who probably hates all white men. At that moment I would have preferred to have been strapped to the outside of a Moon rocket than this. The others sped off.
"You give me the dollar now."
"No have dollar."
"Why you no have dollar?"
"Cos I have spent it."

"You have the cigarettes, whisky?" I was waiting for: 'And wild, wild women?'

"No have." He was certainly driving me 'crazy and insane' as the song goes. I noticed a twitch in his eye and hoped that the twitch didn't migrate to his trigger finger. It's all very well for so-called governments who always say, "Don't do any deals with kidnappers for hostages." They do not have to look down the barrel of a loaded machine gun, held by a mouthy, agitated, inebriated, or drugged individual. I was hostage and negotiator. "Like a Benson?" I offered him a cigarette, and of course he took the whole packet.

"Give 'im owt?"

"No, not a bloody thing, phew."

I got back to my room, poured myself a night cap and sat outside on the veranda, looking down at the Atlantic rollers. Welcome to Africa.

The owner of the oil company arrived from Geneva in his private jet, accompanied by his entourage. It was my first weekend in Cotonou and I was in the lobby of the hotel, when I bumped into, correction, was stopped by, two of his bodyguards. One was, I was to find out, a French ex-karate champion and the other was a Hud from the Bronx, New York. Both were wearing suits, with waistcoats. The bulky kind. I shook hands with the Hud, who immediately withdrew wagging his trigger finger. I have a firm handshake but not a vice-like grip that some have. I once met a bloke in Abu Dhabi with such a handshake. He was from cider apple country 'rrrrrr'. You could actually hear your bones crunch as you shook hands with him and he seemed to take great joy in your discomfort. You virtually had to kick him in the groin to be liberated from his grip. The Hud and the Champ did have their good points, though, like when we ventured into town, we would be sat at the bar having a drink and you could hear smacks and thuds followed by yelps and screams. They used to stop all the hawkers and ladies of the night from entering the bar and pestering us and generally watched our backs, which was much appreciated in a bar in West Africa.

I had been invited to dinner at our hotel by two sales managers from an engineering company, who were trying to get in on the new venture and sell us their wares. We tucked into a meal of frogs' legs and oysters for starters, accompanied by three different wines, then into town for a night cap. I returned to the hotel feeling rather good - well, pissed. I had been wearing these expensive white leather slip-on shoes, which are not ideal for a wet Africa, and what did I do? Yes, I left them outside my hotel room door to be cleaned. And cleaned they were, clean gone. Another African learning curve.

Happy hour at the hotel was always a good laugh, especially when we were joined by one of our accountants, a cockney gay who had everybody in stitches with inter alia, his innuendos. The other bean counter also had his own peculiarities, such as having mouthfuls of water from the local lake. He reckoned that it built up one's antibodies. The lake was situated in the centre of town and, as you can imagine, was used as a toilet and for washing, bathing and swimming, for the locals of course. It didn't do the accountant much good as he spent most of his time on the loo counting toilet paper. One of the Americans, the right-hand man of Dr Minestrone (not his real name either), used to get all worked up and angry if there wasn't any popcorn on his table. He would shout and bawl at the waiters and throw peanuts at them to get his bloody popcorn. When it arrived, I made a point of not eating any. Again, for obvious reasons. You could see the contempt in the waiters' eyes, and who could blame him?

After I had advised the technicians on their days work and attended the morning meeting, I would grab a cup of coffee and go into John the safety/security manager's office for a chat and to bring each other up-to-date.

"Guess what I did last night?"

"Well, I guess you are going to tell me."

"You know that little shit, the Yank, the second in command who throws a tantrum when there isn't any popcorn on his table at happy hour?"

"Yeah?"

"He had gone into our main office a couple of nights ago. Of course my security guys knew who he was and didn't stop or question him. He created a big stink over not being challenged and tried to drop me in it."

"Go on."

"So, I broke into Dr Minestrone's villa last night…"

"You did what?"

"The guard was asleep. I took the keys from his belt, got inside and quaffed a few glasses of his nib's champagne, then I put yellow stickers on all the bedroom doors with 'Bang you're dead' on them."

"Bloody hell."

"Can you imagine the scene the next day when they saw the post-its? Well, the little shit has been fired, along with Pinkie and Perky, the Hud and the Champ, and guess what?"

"There's more?"

"He found out who did it."

"How?"

"Because I told him. He's now offered me the head of his security position."

"Are you going to take it?"

"Nah, couldn't be into all that crap." John had been in the forces but he wouldn't say with which branch but I could guess and I think I'd be right!

I paid one of my weekly offshore visits to the oil rigs. The choppers were grounded due to lack of spare parts, so I had to go by boat. To transfer from a rig to a boat and visa versa one had to ride in what's called a 'Billy-Pugh', which is basically a bell-shaped sort of rope ladder, with a round bottom. When you first use one, you think that you go inside but no, you have to stand on the outside and hold on, like a white-knuckle ride, which can be quite nerve-racking as they go high. Lose your grip and you are a candidate for the pearly gates or stoking fires! Thank goodness these days one sits in a transfer pod called a 'frog' and are strapped in. The ocean this day was quite choppy, the boat was pitching all over the place and the crane driver mistimed our landing. I bounced off and started to roll towards the stern of the ship. As they do not have handrails at the rear, if you went overboard you'd fall straight into the propellers and end up in a bit of a mess. Fortunately the ship pitched the other way just in time. 'A life on the ocean waves', hey?

We were into happy hour and I was waiting to tell my tale of the day, when in walked the Rothmans cigarette rep, a fellow Scouser and Everton supporter. He was as white as a ghost.
"What's wrong with you?" I asked.
"I was driving along the main drag, when that fucking witch doctor ran into the front of me car."
"Did you hit him?"
"No."
"Why not?" The caring attitude of fellow ex-Pats.
"He was staring at me in a trance-like state and put this six-inch nail through his nose. Scared the shit out of me." It was the best story and best laugh of the day.

The following weekend we had the misfortune of encountering this nail-fan witch doctor. It was a Saturday afternoon and we were lounging round the pool having a beer – well, the gay accountants were, the rest of us would be spilling our beers down our chins, due to the fact that the Norwegian women, who were already topless, would get out of the swimming pool to take a shower and in a typical Norwegian way, oblivious to us, they would semi-remove their bikini bottoms. This particular Saturday, we had all been advised not to go into town as there were going to be heated demonstrations about the government's latest 'screw the populous' policies, which usually end up with the demonstrators being man-handled, jailed or shot. John, the security manager, said, "Stuff this, I'm bored, let's go into town."

We piled into several taxies and headed down the main drag to our favourite bar. It is advisable in these places never to get a seat with your back to the door. I got in quick, followed by John, and ordered a round of drinks.

"What the fuck is that silly c*nt doing?" John said. The bloody witch doctor was standing at the door and one of our gay cavalier accountants was inviting him in to join us - 'oh shit.' A Scottish guy and his wife whom we had got to know had joined us. He had lived in Benin for over twenty years and had married a local woman. His wife had told us that she was related to President Mathieu Kérékou, which we had kind of taken tongue in cheek. She stood up, went over to Doc and gave him a mouthful. What she said I have no idea, but Doc said something like:

"Oo ee oo ah ah, bing bang walah walah bing bang, oo ee oo ah ah bing bang walah walha bing Bang," as David Seville's 'Witch Doctor' song goes, and vanished. Not literally. Which was a disappointment. I do believe that she had been telling us the truth after all. Who else would have the power to scare the shit out of a witch doctor but a relative of Mathieu K, the president?

We were all invited, or to put it in another way, told, to attend a party given by Dr Minestrone at his villa. Dr M was stood at the head of his hoodlums, who were all dressed in the obligatory black suits, with waistcoats. They stood facing each other as if in a wedding archway, where they hold swords above the bride and groom. I was the last to be introduced to the Doc. He asked questions such as: 'How were things generally going?' and 'Did I like the job?' He must have known who I was anyway. We chatted for what seemed like a lifetime but could have only been a minute or two and as I walked away, I stopped, turned around and said, "Dr Minestrone, if you ever need a job, come and see me and I will fix you up."

There was silence. 'Oh fuck.' What was I doing, saying that to him, the owner of the oil company and (I suspected) possible Mafia boss too and in front of his hoodlums. With that, he started to laugh, which kind of, gave his Huds permission to laugh too. I must have hit the right chord, as, on some occasions when I'd been into town, I'd sometimes have a night cap in the hotel lounge bar and one night Dr Minestrone was there with two of Benin's oil ministers, who must have been boring him to death. He saw me and called me over to join them. We started chatting about football and motor racing, his beloved AC Milan and Ferrari. I was in his good books. Well, one hoped.

At our morning meetings, there were in attendance two American World Bank officials as observers. I got on well with both of them and used to give them both a lift from the hotel to the office. One of them said to me after a meeting, "I don't like the way these guys do business." What could he mean? He told me that he suspected (as I had) that 'these guys' were the Mob, as in Mafia, providing a cleaning service to the local industries - a money-laundering business, what else? Somebody was being taken to the cleaners, put through the mangle and hung out to dry. And this is the reason I changed the names of the company and the boss!

The hotel threw a party for the American Navy, as one of their ships had just arrived in Cotonou, which just happened to coincide with the arrival of a Russian dancing troupe that was in town at the same time and who were also invited to the party. I was outside on the veranda having a drink when I was approached by a Russian, who spoke perfect English. He started asking me some loaded questions about the oil fields, such as production rates, engineering problems and budgets. I told him that I was here to enjoy myself and didn't answer any of his questions. He politely excused himself and got amongst the Americans. 'Now I wonder who he was?' Rumours were flying around everywhere; it was beginning to be intriguing. A report out of another old newspaper cutting, also found by the bean counters, was a story about a plane landing in Cotonou. A bunch of mercenaries disembarked, then made their way into town, shooting and doing what mercenaries do, like kill people. On their return to the aircraft, they popped into a bar for a beer, as you do! Then they flew out, as if they had been on a day trip to Alton Towers. Now this gave the President some street cred on how he had risked life and limb and fought off this marauding army who had almost met their nemesis to protect his people. Yeah, and pigs might fly.

"He took you down what road?"

The rookie was at it again. My driver, who happened to be twice as bad as Maureen from 'Driving School' who had failed her test over 100 times. He was 50% blind and 100% dim, not a good combination for a driver. He had driven me past the Presidential Palace.

"Christ, Ken, have you seen Paul's car?" Paul worked for another company in Cotonou.

"No, why?"

"It's full of bullet holes."

Paul had driven down the same road, only a couple of days ago and was shot at by palace guards. His car was riddled with bullets. Miraculously he escaped without stopping one but had to change his underpants and had almost had a breakdown, a nervous one! I ended up having to dismiss the driver before one of us was killed or badly injured. In a country that beats you to death on the spot if you run over, kill or maim one of their family members, it is not only the driver that gets it. Scary.

My office was the open type, meaning it had gaps at the top of the walls, which was a damn nuisance. When you opened the door in the morning the place would be full of mosquitoes. You had to duck to get to the air-conditioning unit to switch it on. If there is one thing that I despise most it has to be the mosquito. Some mornings you would wake up to find that you have three foreheads and look like John Merrick, the Elephant Man and they are always behind you in the mornings on the wall, full of your blood. You splatter them with your shoe and the wall would make a masterpiece for the Tate Gallery, red splodges everywhere. The Saatchi brothers would pay a fortune for it. The air conditioning soon takes its toll and they drop like, yep, flies. Hypothermic mozzies.

In fact, they are the biggest killer in the world. The times I've spent worrying when I sneezed or got the sniffles after being bitten. When you read the information on the packets of prophylactic tablets, you begin to wonder if they are they worth taking. (They most certainly are.) Some play havoc with your eyesight and your kidneys and the very latest ones, the one-per-week type, Larium, you can go absolutely bonkers on those: you become anxious and worry about everything when you haven't got anything to worry about. You worry because you think you've forgotten something that needs to be worried about. You can also hallucinate. I could swear that I saw a pink elephant on one African trip, or maybe it was the G&Ts or the DTs, one will never know. You cannot afford not to take the tablets, as they are terribly important, especially in Africa where there are different strains of malaria to each area. So be warned, read up on it and get the ones that suit you best. Better well read than, well, dead!

I left Africa more knowledgeable. I could tell the difference between a genuine ebony statue and a white wooden one that had been boot-polished, I could spot fake ivory a mile away, I knew which streets not to go down without being shot at.

I could also deal with witch doctors (or at least connect you with a woman who can). But seriously, I have travelled quite a bit now in Africa. It is wild, beautiful and can be dangerous, a very exciting continent with a new experience every day - not to be missed and messed with at your peril!

LIBYA, 2ND TRIP 1986

(In which I meet with cockroaches, pongs and a moving wall of sand.)

"When are you flying out, Ken?" It was my new boss.

"Monday." Which was in two days' time.

"Why don't you wait for me? I only have ten more days' leave."

"I don't mind going alone really, thanks all the same."

"Suit yourself, I'll see you when I get there."

I was eager to get going and didn't want to lose the pay. I had to stop off at Malta en route to collect my visa to get me into Libya, as their embassy in London had been closed due to the killing of WPC Yvonne Fletcher, the policewoman gunned down during a demonstration outside the Libyan embassy in St James's Square.

I'd been to Malta several times before and it was no hardship, especially now it was summer. The next day, I arrived early at the Libyan embassy so I could catch the afternoon desert flight. I rang the doorbell several times until eventually the front door slowly opened and out popped a head that had not seen a comb or razor for several days.

"What do you want?"

"I've come to collect my visa."

"Why?"

"To enable me to enter your country."

"For what?"

"So that I can work there." Of course it wasn't ready, what did I expect?

"Come back bukra." (Tomorrow)

Shit. It would have been a waste of time arguing and no point in stretching the issue. Another night in Malta would not harm me. Besides, he was using the IBM system: (**I**) Inshalla/God willing, (**B**) Bukra/tomorrow and (**M**) Malish/never mind. The next day, armed with my visa, I headed to the airport to check in. A swarthy Libyan chap approached me:

"Can you take this box to Tripoli?" He was holding out a sealed box, containing, so he claimed, a tin of paint.

"Er, no," I said.

I later heard this was a common thing but who in their right mind would say yes?

Arriving at the office in Tripoli, I met the manager Frank Grivas (not his real name), aka Maltese Frank. It was he who had persuaded me to come to Libya. I was yet to find out he was of the mendacious, duplicitous type of Homo sapiens. Basically, a c*nt. Frank was busy on the telephone roasting somebody; talk about Sir Alex Ferguson's famous hair-dryer treatment – this was a hurricane. What a temper. I was then introduced to an even more swarthy Libyan and asked for my passport.

"What for?"

"For safety."

"It's safe enough with me."

"Company rules," I was told.

I didn't like this one bit. It had happened to me before in other Middle East countries. They take your passport off you and you don't see it again until you're leaving the country. This always worried me. What if you had a problem at home and you had to get out that day? In addition, what if it was the weekend? It states in a British passport that you should not pass it over to an unauthorised person. Again, what can you say? Then you can have problems of getting an exit visa. This can be a heart-stopping time. In Iraq, while Saddam the Sadist was in power, they had me wait a whole day, then some git would give you the third degree. They must be hand-picked for the job.

Eventually, I was taken to the refinery where I met some ex-pats. One was from Liverpool. We had a brief conversation, then he showed me around the place.
"Well, I think I'll go and find my room and unpack before dinner."
"You're not staying here. I'll get you a driver, he'll take you to the camp."
"Camp, what camp?"
"It's a contractor camp on the outskirts of town."
My heart sank. What is this going to be like? I had to leave the close confines of Marsa el-Brega, where all the ex-pats were living, to go God knows where? My heart sank even deeper when I set eyes on the place. It looked like a shanty town and I was the only Westerner there. My room consisted of a bed, a small wooden kitchen chair and a desk and wardrobe made out of packing cases. On the wall was a photograph of Little & Large the comedy duo. The randomness of that actually cheered me up a bit. I was later to find out that Sid Little was the son-in-law of my new boss, hence the photo.

That night I walked to the gate feeling very low, when I spotted a familiar face from the past across the road. It was a South American guy I knew, who reminded me of the band leader Edmundo Ross. I'd worked with him in Basra, Iraq, when the Iran/Iraq war broke out, but that's another story. We were both lucky to be alive, even though we were in Libya. After catching up with each other, he told me that a Brit was in the same camp as himself and would enjoy some company. We were introduced and spent a good few hours sampling home brew and telling 'When I Was In…' stories, about past contracts.

Then, disaster, my job fell through at the refinery and I had to stay at the shanty town for another week until my boss arrived. Maltese Frank asked me to carry out an inventory of spare parts for the equipment at Marsa el-Brega, then drive to Benghazi to bring some parts back. This I was glad to do, as I was bored sick of where I was. A local driver and I drove across the desert and arrived at the company guest house quite late. The driver and I were starving, as we had set out with only a coffee for breakfast and had nothing to eat all day. We ate a plate of fish with rice, followed by a cup of sweet tea. No such thing as one or two spoons; it's one or two bowls but better than nothing. Just. Strange thing, you never see dentists in these places.

I was shown to a room full of multi-ethnic lorry drivers. It stank of curry, farts, smelly feet, bad breath, underwear that had not seen the light of day, let alone a washing machine for at least six months, BO, stale cigarettes, and to top that, no air conditioning. The mattress had been dragged through a minefield and shat on by a herd of cattle and there was no bedding. Absolutely no way was I going to spend the night there. I went onto the flat roof for some air and bumped into a pair of crocodile-skin boots, a Stetson, huge belt and cigarette smoke. This was of course an American oil man. Was I glad to see him.
"Gad-damn Libyans."
"What's wrong?"
"Those motherfuckers haven't paid me for three months and am gonna to stay right here until they do."
"How long have you been here?"
"Three god-damn weeks."
"What, here in this motherfucking guest house?" (Americanisms creeping in here.)
"Sure, here with these assholes on this motherfucking roof."
"What did you sleep on?"

"One of them beds in the corner there."

"Without a mattress?"

"What you expecting, the fuckin Ritz?"

"Not really," I said, sounding like a real plonker. "Mind if I use the other bed?"

"Help yo self buddy, better than sleeping with them motherfucking dirty bastards, smell like shit. I'd shoot the whole gad-damn lot of em. Name's Hank, from Houston, how bout you?" "Name's Ken from Liverpool, England."

"No shit, a Limey hey?"

"Yep."

"Gad-damn Limey. You know them Beatles?"

"Sure, know em all."

I had never even set foot inside the famous Cavern Club in Liverpool. I must be the only Scouser of my age that has never even seen them live.

"Well, son of a bitch. Better hit the hay as they say. Night, Ken."

"Night, Hank."

I lay there on springs, using my travel bag for a pillow and my towel for a blanket, looking up at the stars, wondering what I'd got myself into, yet again.

After one of the most uncomfortable nights ever, as I don't usually sleep on bed springs, I sprung into action. I went down to a breakfast of ice-cold, freshly squeezed orange juice, bacon, eggs, tomatoes, baked beans, hot buttered toast, fried bread and hot filtered coffee. ....Well no, it was cold fish and rice followed by, of course, sweet tea. This is where the expression 'beggars can't be choosers' comes from.

"See you next time, Hank."

"You take care of yourself, buddy."

"Hope you get your money."

"Sure kick their asses if I don't get it today, motherfuckers."

It is amazing the amount of people from all over the world that you meet in this business who you never see again

- sometimes good, sometimes sad. Surprisingly, the week passed quickly and I had to go to the airport to pick up my boss at Marsa el-Brega. He was only staying overnight, as the next day we had to drive across the desert again in an old Land Rover. Just as we got off the black-top/tarmac road, we stopped. The driver started to deflate the tyres for desert driving, as we didn't have the proper ones on the vehicle. 'Well, that's a good start,' I thought. Bill, my boss, started to have a pee. "Have a pee now," he advised, "as you don't pee too much in the desert."

"Oh really?" I said, desperately trying to write my name in the sand but only got as far as the 'K' with full stops.

After several hot, sweaty, bumpy miles without air conditioning, driving over sand dunes and past Wadis/valleys, almost getting stuck, we arrived at the camp. The crew consisted mainly of Arabs and Africans and they greeted us warmly, as they obviously thought a lot of their boss. The living quarters were sparse. I was to share a room with an Algerian welder who happened to be on leave, which was good. The middle of the double Portakabin was a shower that stank of rotten eggs. H2S: hydrogen sulphide. H2S is a sour gas that is a colourless, flammable compound with a distinctive 'rotten egg' smell. Gas containing even low concentrations of H2S is toxic to humans and animals. A concentration of 10ppm (parts per million) is the threshold limit, or the concentration in which a person can stay for eight hours without problems (though I'd not like to: eight seconds is enough!) At 100 ppm, the sense of smell is deadened in, say, three to fifteen minutes. This can lull you into thinking there's no longer any gas, when in reality the concentration could have increased. So using one's nose is unreliable and could even be fatal. Above 100 ppm, respiratory problems and paralysis set in quickly and, unless the victim is rescued immediately, they will die. At 1000 ppm and above, the victim will become immediately unconscious and their breathing will shortly stop. At 5000 ppm, it's immediate death. Not nice stuff to mess with!

The water came up straight from the stinky ground and was 'filtered by pebbles'. 'Jesus, I hope these guys don't smoke,' I thought. On the other side of the cabin, sharing our eggy shower facility were two Syrians, one a civil engineer (who was the image of a young Sean Connery) and his friend, who was an assistant cook. Both turned out to be really nice guys and I got on well with them as I'd been to Syria, which gave us common ground. After our evening meal, we sat outside. The crew gathered around Bill, who told them a tale or two; some very dubious I must add, but enjoyable. He told them one about his next-door neighbour, who had complained about the state of his garden. "What did you do?" "I got the local garden centre to relay the lawn the same day," said Bill. "Should have seen his face, the toffee-nosed prick was sick."

That night, I climbed into bed - my first night in the desert. Not a sound but for the wind whispering through the sand dunes. I started to read a book and, after a few minutes, I thought I saw something move in the room. I put the main light on and shuddered at what I saw...cockroaches climbing up the walls. Oh shit. How could I sleep with that little lot? I killed as many as I could with my shoe but the thought of having to sleep in a mini zoo didn't do my mind any favours. I pulled the bed away from the wall and checked under the mattress and pillow. I then went over to the workshop and brought back some lubricating grease which I put on all the bed legs. Satisfied with what I had done, I got back into bed and opened my book. Minutes later, the same again. Where in God's name are they coming from? I killed some more. The room was awash with dead cockroaches. I ended up having to sleep with the main light on all night like a baby. I could have done with a dummy to stop them crawling into my mouth. As you can imagine, I had a very bad night's sleep.

The following night, I heard this commotion outside. I opened the door and this horrible face was in my horrible face, a bloody camel! A herd had come into the camp and were drinking the water out of the fifty-gallon fire drums that were outside each person's accommodation. I got into my truck and chased the herd out into the desert with my headlights full on. Camels will eat and drink almost anything. I have seen them eating empty cement bags. So I wondered what the third night would be like...

The job that I'd came out to do hadn't materialised. The thing with some overseas contracts is: they can screw you up no end, especially when they hold your passport. So here I was, in the bloody Sahara desert, minus job, minus passport, thinking 'Will I get paid?' just like 'Hank the Yank' on the roof of the guest house. Bad news: Bill had to go back to the UK, meaning I'd be left with this crew for a week, maybe two. Some of the Africans and Arabs used to get fired up after chewing on ganja, which made their eyes bulge and their mouths blood-red like Dracula. The main problem was that they started to pick fights with personnel from another tribe or area. I ended up like the United Nations. Well no, I actually got off my arse and stopped the fighting! Apart from doing unpaid work for the UN, I did some civil-engineering work to complete my first trip. I returned from the desert to go on my field break, only to find out that I was to be transferred to another desert location. At the guest house, I couldn't help hearing a bullshitting Geordie telling someone that he was a 'superintendent'. Little did I know we'd be based at the same location.

Before I left, I was told by the other smooth-talking, bullshitter Frank (Maltese Falcon) Grivas, that when I returned it wouldn't be for the same job I'd signed the contract for. I should have known what kind of person he was when, after a few home brews affected his central nervous system, he let slip that he could not return to Malta, at least not while

Dom Mintoff was in power, as he'd been a 'naughty boy'. To say that 'being pissed off was an understatement', is an understatement. Being underemployed, to twiddle one's thumbs all day, can be a pain. I was on a six-weeks-on/three-weeks-off rota and stuck out in the desert can be a long tour of duty, especially when there's nothing to do and no place to go in the evenings. The thought of having to go into that office full of swarthy Libyans and Maltese Frank (who was, to put it mildly, a fucking crook) and tell them to stick their job in a place that sees very little sunlight did not make me go to work with a spring in my step. But it had to be done. I walked into Mad Frankie's office and put my cards on the table, expecting the Fergie hair dryer, and was surprised at his reaction. "Don't worry, Ken, I have the perfect job for you," which made me worry anyway. "What's that then?"

"Transport manager."

"Correct me if I'm wrong but you only have about three vehicles in the whole of the Sahara desert. What is there to manage?"

"Our company is expanding and we're thinking of buying some of the French oil company's trucks."

"What kind of trucks?"

"Oil rig moving trucks."

"You mean those huge Kenworths with an engine like a train that have about 20 gears and huge wheels?"

"They're the ones and I want you to check them over and do a survey on them with a view to buying and setting up a maintenance depot with your own staff."

I returned to Libya with a fresh heart. At least I could make this job interesting.

En route, I had to change planes at Geneva, from Swiss Air to Libyan Arab - quite a contrast. While walking towards the plane, I noticed all the baggage was on the tarmac besides the aircraft. Unbeknown to me, you were supposed to load your own bags onto the plane. At Tripoli airport, I waited and waited by the carrousel. No sign of my baggage. It would have been a terrific compliment to call the staff "a bunch of bone-idle bastards," but one showed a bit of interest by yawning and picking his nose. "What your name?" When I told him, he yawned again and said, "Come back, let me guess, bukra." So it was another night at Mad Frankie's guest house and who should be there but Hank the Yank, still waiting to be paid.

"How goes it, old buddy?"

I sprang into Americanisms again. "Would you believe those motherfuckers have lost my baggage?"

"No kiddin, Ken, don't hold your breath, doubt you'll see that again. I lose one every other gad-damn trip. I brung shit all this time: hand-carry cigarettes, got two pair of jeans and a couple of sweat shirts are on the rig."

I thought of all the goodies that I had brung with me. The next day, I returned to the airport. "All lost baggage goes in the hut outside," I was told. A guard took me to the hut, which turned out to be an old airplane hanger that could house about four jumbo jets. I immediately lost all hope as I looked inside. There must have been a million bags and suitcases all piled to the roof.

"Which ones came last night and today?" I asked. The storekeeper laughed, burped and shrugged his shoulders. "Do you keep any records like airlines, flight numbers, dates, times, the countries that they came from?" He looked at me, vacant and disinterested. There was no order whatsoever, just a massive pile of people's belongings right up to the roof, including mine. And by the look of it, some had been there since Rommel and Montgomery were sparing partners. I started the climb, as methodically as I could, and ended up like Eli Wallach's Mexican bandit in 'The Good, the Bad and the Ugly', running around like a headless chicken looking for one name on a thousand headstones.... I gave up. Two days later, I phoned Swiss Air and was told:

"Yes, we have your baggage, or what is left of it."

"What do you mean, what's left of it?"
"Sorry, sir, but we had to blow it up, a controlled detonation."
"You had to... what?"
"You'd left it behind on the tarmac and with it being Libyan Arab Airlines, we were suspicious and had to send for the burm squad thinking that it was a burm. We had no choice but to make a controlled detonation! Sorry, sir, but you should have put it on the plane yourself."
"Nobody at Swiss Air told me that!"
"Sorry, sir`," the disinterested robot airline voice repeated.
My poor Breville toaster, my new Marco Polo sweater... were now flying over the Alps, like Swiss cheese, full of holes.

I flew into Naphora in the Sahara, which was to be my home for the next six weeks, and arrived at the camp. Before I put my one small bag down, I could hear a high-pitched Geordie voice screaming orders to some poor African who was probably earning $20 a month and got home once a year, if he was lucky. I like and get on well with Geordies but this guy, whom I'd met before in Tripoli, really grated on me. He was a jumped-up little prick, a welder who in his mind had risen to the dizzy heights of superintendent - in Libya for God's sake - where he could boss all the poor Africans. I can imagine what would happen to him in a shipyard back in Merseyside... This had obviously gone to his small head. I kept out of his way as I just knew we'd clash. I got that feeling when I first set eyes on him in Mad Frankie's guest house in Tripoli.

I started to set up my new transport department, which incorporated the garage and plant vehicles, i.e. earth-moving, pipe-laying machines, compressors generators and welding machines. The following week, two ex-pat mechanics turned up, who were to work for me: the first clash. We were busy and had the day's work planned, when the demented Geordie came to the garage and started to give orders to the mechanics. I thought, 'Here we go.'

"What is it you want?"

"Just given the lads something to dee," he said in Geordie.

"Sorry but they're busy."

"I'm the superintendent," he said.

"It appears that we both are, obviously of different departments. Why don't you go back to your department and I'll take care of mine."

I could feel the hatred inside him. He stormed off, not knowing what to say or dee. I had to nip it in the bud straight away, as it can't be good, mechanics taking orders from two people.

The following day, Joe, one of the mechanics, and I set off early to drive to the French oil company to inspect some oil rig moving equipment. Joe told me that he used to be a gunrunner in Africa and was used to roughing it, which was just as well because that's what we were doing out here. "Your not running anything here, are you?" I asked. He nodded his head in the direction that I wanted him to. We inspected the machinery and got it fired up after a few adjustments. Apart from a few minor repair jobs, it was in decent working order. We were staying overnight in the company's guest house which was 'rough' to say the least. That night, I had to drive over to the oil company to attend a meeting. The distance was about two miles. 'No problem,' I thought as I could see the lights. As I approached the gates, I noticed a Libyan guard stand up and point his rifle at me. I was about fifteen metres from him. 'What the..?' I slowed down; he slotted a bullet down the spout and aimed at me. I slammed on the brakes and froze. He walked over and pointed at the headlights.

"Whats wrong?"

"Lights, lights."

I had my headlights on and they were dipped, as I didn't want to blind him as I approached the gates. "Lights, lights," he kept on repeating. What the arse meant was, I should not have approached with my main beam lights on, if that makes sense.

"I have low lights on, no full!

"How I see?" There was no way I could get into his bird brain.

"Very sorry, no do again." What else can you say to an idiot pointing a loaded gun at you? I handed him my desert pass, rotating it the correct way for him. Not a nice experience, looking down the business end of a cocked rifle, in the hands of a trigger-happy Libyan soldier but an experience it was.

The next day, Joe and I headed back to camp. We'd been given what looked like a prospector's map. It seemed like a joke as it had oases, wadis, oil fields, drilling rigs, columns of smoke and something that made my heart miss several beats: minefields, a legacy from the Second World War. "We'll be alright, just following the tyre tracks in the sand," Joe said. Mistake number one. We bombed along. (I think we 'drove along' would have been a more apt term in our circumstances.) It didn't seem to be a problem, until we came to a desert crossroads, where hundreds of vehicles had passed, intermingling the tracks to such a degree that you couldn't tell which way to go. We tried to backtrack the way we'd come but soon got completely lost and just didn't know which way to go. I stopped the truck to contemplate what to do next and have a drink of water. Mistake number two. To my horror, our water container had tipped over and had a crack in it. We had only an inch of water left and this now had sand in it. We hadn't tied it down enough and were lost in the desert, just like the movies.

It was August and the temperature was over 100 degrees F. The biggest problem, however, was below our feet - the mines. "Well, Joe, I reckon we could be in deep shit." He started to laugh but I failed to see the funny side at all. We obviously couldn't stay due to not having any water. Should we sit it out like you're supposed to do in this situation and wait for the cavalry? Nobody was expecting us, so wouldn't be sending out a search party. We didn't even have a radio. If we left it until dark, we stood the chance of going over a sand dune that had collapsed on one side like a cliff edge. It was bad enough to travel during the day, as the sand immediately in front of you blends into the sand miles ahead. You could be driving along a dune, thinking it was continuous, then come to where it had collapsed and fallen away and over you go. Not a nice experience as you can well imagine! And of course there's no emergency 999 service out in the desert. That's if you were still alive. Our only method of navigation was the Sun. We observed it for a while to see where it was setting and roughly worked out where the Mediterranean was as we needed to head that way and not deeper into the desert. The feeling of driving along, not knowing if you were going to be blown to smithereens at any moment, was not my idea of fun. To our great relief, we spotted the main black-top road going into Marsa el-Brega. We knew the way to our camp from there, so I lived to tell another tale.

The following week, we carried out a full survey on the huge land oil rig moving vehicles and gave the go-ahead on two. The next thing was to bring them back. I couldn't help but feel a wee bit apprehensive at the thought of driving one of these monsters. Each wheel was about five foot diameter, twelve to each vehicle. To climb into the cab was like climbing a rock face and sitting behind the steering wheel was like looking out of the bridge of a ship. It had, I think, twenty gears. I obviously didn't hold a licence to drive these beasts and being in Libya without one didn't quite fill me with joy. After a quick lesson from Joe, who was a heavy vehicle mechanic and had driven one of these before, we set off, him in front. Selecting gears was like putting your hand into a jar of jelly beans, trying to locate a grain of rice. I was just getting used to driving the monster when I looked up in horror at the sky. One of the most awesome sights that I have ever seen, a sand storm, was heading our way. It was like a huge black wave that was half a mile to a mile high! And it was heading straight for us. 'Oh fuck.'

I blasted the huge horn at Joe for him to stop. "What do you reckon?"
"Ah, no problem," he said. "Just keep your headlights on."
"Wouldn't it be advisable to sit it out until it passed?"
"Nah, just keep close together."
"Well, I hope you're right." I climbed back into the truck, closed the windows and vents and did a Lawrence of Arabia, by putting a chequered Arab scarf around my head and face. I watched, as it got closer and closer. One second you could see, the next, nothing, not even Joe's huge tail lights. I kept going for about five minutes, then I had to stop, as it would have been sheer madness to continue. What if I knocked down an electric pylon or crashed through a main oil pipe line or even worse, ploughed through a village, as there were many Bedouins all over the Sahara. I decided to sit it out. Even with the windows closed and old rags covering the vents, the sand still got through. I could smell it, taste it and for one horrid moment I thought that I would suffocate as it was now getting very stuffy inside the cab. The storm lasted for well over an hour. I peered out of the cab and could see our camp. I shuddered to think what could have happened, if I hadn't stopped. But where was Joe? He showed up about two hours later. He'd gone way past the camp and had been going round in circles. What did he expect? He was bloody lucky!

Safety in the oil and gas industry is paramount and rightly so. Who could forget the Piper Alpha disaster? I've worked for some of the world's largest oil companies – Shell, BP, Exxon Mobil, Total etc - and safety is instilled into you so much that it comes out of your ears. Every job that's carried out on an oil rig or an oil refinery etc, demands that you must have a correct work permit and wear the right PPE (personal protective equipment) and it still may not be accepted and then you'd have to do it all again. The permits are legal documents and are kept for about ten years. You could have several technicians and engineers unable to start work, hanging around and this would not go down to well with an O.I.M. (offshore installation manager) and the oil company. So, being used to such a stringent system as this, you can imagine the shock to my system when I found myself in Libya, in the Sahara Desert, working for a Libyan oil company where safety is non-existent and on many occasions they'd have you drive alone into the desert, in a clapped-out vehicle, minus a radio. One of my main concerns was getting bogged down in sand or getting lost. There have been many fatalities where people get stuck, run out of water, then abandon their vehicles and walk. Even though they've logged themselves out of their base, it's very easy to become disorientated as everywhere can look the same and you could end up walking around in circles. One day, you could be following tracks or certain sand dunes, then a shamal/sandstorm would come and hey presto, no tracks, no dunes and your map makes no sense.

A few years later, I was working again for Shell and had the excellent opportunity of being taught desert survival by a team of British ex-soldiers that had spent some time in the Omani Desert. They wouldn't exactly tell you what they'd been up to, but it wouldn't take a brain surgeon to guess which regiment they had belonged to. Some of the course was rather hairy, as they would get you to drive fast in the sand and then make a sharp turn, this was so you could get yourself out of a rollover. They showed you how to collect water and many other things to survive. On future trips to the desert, I would take an ample supply of water (this time fastened down), food, matches, a cigarette lighter to set fire to the spare wheel to make smoke which would hopefully be seen by a search-and-rescue party, some sacks to go under your wheels to give you grip, various tools and a spare fan belt. Also, I would log myself out, give the type of vehicle I was in and never carry extra cans of fuel because, if a spotter plane was looking for you, they'd not be able to calculate how far that type of vehicle would go, as they would not know how much extra fuel you had on board.

I had mixed feelings about Libya. I liked the desert – I may have got it from my dad, who was a Desert Rat with Monty in the Second World War. I wonder if he had laid those bloody mines! I vowed that I would not return working for a cowboy outfit such as Mad Frankie's but hey, this is how you gain experience in life, sometimes easy, mainly hard, but you learn! Or do you? Oh, I did get paid. Sorry, Hank, wherever you are. Another oh: I didn't return to that dreadful company again. IBM.

IRAN, 1ST TRIP 1987

(A health-and-safety-free zone.)

I was in the departure lounge at Frankfurt airport and couldn't help but notice this huge guy being harassed, cursed and chased by a petite woman. She was giving him some stick. The guy was built like a brick outside toilet and his face seemed familiar. It was Floyd Patterson, the ex-world heavyweight boxing champion. He must have felt a right Charlie, as nearly everyone had recognised him. Some people tried to get his autograph!! It was like trying to stop a runaway train because he was certainly blowing and was obviously in a foul mood. I dived under a table out of the way. Well, no I didn't but if he'd headed in my direction, I wouldn't have stopped him under any circumstances! I was still grinning when I boarded my Lufthansa flight bound for Frankfurt, en route to Iran. The steward had just served breakfast and I looked at the front page of my newspaper and couldn't believe, or rather didn't want to believe, the headlines. The foreign secretary was calling Iran 'the gangster state' and very much more and I was heading there. Don't they realise some people have to visit these places?! This had happened to me before, en route for Moscow when the then U.S. President Ronald Reagan called Russia 'the evil empire'. That wiped the smile off my face.

I arrived in Tehran somewhat nervous, as I remembered my outburst when being interviewed on TV at Schiphol Airport,

Amsterdam, when I slighted the Iranians something awful after they'd bombed a petro-chemical plant in Iraq that I was working on and killed some of my colleagues. I had mixed feelings of sadness and anger, both anxious and happy to be alive. This was on the BBC six o'clock news, which more than likely would have been relayed to Iran. I was half expecting somebody to tap me on the shoulder and say: "We remember you, step this way." The company's guest house was in the centre of Tehran, where I spent a lonely night as the only one there. Every company guest house in the Middle East is the same, over the top and crammed full of chandeliers and cheap fake antique furniture that never gets used. The following day, a company driver in a huge American car picked me up, which was weird enough. Driving past huge placards of Ayatollah Khomeini made the hairs on the back of my head stand up somewhat. The driver started his questions:

"Where you from?"

"UK."

"You British?"

"Yes. What a beautiful building, is it a mosque?"

"Which part British you from?"

I sighed inwardly. "Liverpool."

"Ah, Liverpool football team!"

"I support Everton, the opposing team." That shut him up. There are only two teams anyone abroad will ever talk about, Manchester United and bloody Liverpool. It's always tiresome having to converse with someone after a long flight and not much sleep, when you don't speak the language, especially when they say the words "Liverpool football team." I tried a new tack, saying 'Everton' as often as possible.

"How long you be here?"

"Everton."

Pause.

"You have children?"

"Everton."

Pause.

"You engineer?"
"Everton."
Pause. I was being a prick but he got the message.

We arrived at the Japanese Gas Corporations office and I was introduced to everybody. All the bowing made me feel like royalty. With it being a Japanese company, the handshake was always accompanied with a bow, even by their project director. I just couldn't do it and made a point of not doing it. All morning was spent drinking cup after cup of coffee whilst waiting for something to happen. After lunch we set off to the site, which I was told would take about three hours. I was quite surprised with the place as it was more modern than I expected and clean, then we drove past the ancient religious city of Qom, where Ayatollah Khomeini's son and "right hand", Ahmad, was born. Nearly every building was a mosque. Well, I suppose it is a very religious place, a holy city. About halfway, we stopped for a bite to eat in what I suppose was their equivalent of a transport caff in UK. I felt hundreds of eyes looking at me and whispered conversations started up. What does one eat in a place like this? Better not ask for a bacon sarnie! I didn't like the look of some of the meat that some of diners were eating so I spotted some feta cheese and tomatoes. That and some pita bread will do me.
"You no want meat?"
"No, don't feel too hungry."
"You must have the meat."
"I don't want any meat." I was getting exasperated with this guy. "Everton." Silence at last.

We set off again. It was too hot and the air conditioning wasn't working too well. Then we came to a crossroads where a car was parked. My driver stopped, got out and went over to the other car. The other driver got out to greet him. They shook hands, kissed each other in the usual Middle Eastern way, then appeared to argue over something. This went on for some time and I was beginning to feel a little uneasy about the situation. My driver got back into the car without saying a word and drove off the main road and into a side street. Hmmm.

"Where are we going?"

"House, house." The once-friendly driver appeared to have changed. I was supposed to be going to the site office for a meeting, not to a house in God knows where. He just drove on down the back streets. I'd never find my way out of here in a month of Sundays! Alarm bells started ringing as I thought of the hostages, Terry Waite, John McCarthy and Brian Keenan. To say I was alarmed and worried would not have been an over exaggeration but why me? I'm just an engineer. We soon stopped at a house.

"Come, come."

"No. You take me site, I want to see Japan man." Christ, I was talking like him now. I had to gather my distressed state of mind and think. Was I to become a hostage? I was in a little village somewhere inside Iran, with a psycho driver. The UK didn't even have an embassy here and I was working for a Japanese company. 'Great.' All my fears about coming to Iran were now starting to unfold. Why oh why did I come here? Well, I suppose the salary helped somewhat but who's thinking of money in a situation like this? After a small stand-off, the Liverpool fan and I got back into the car and drove off into the mountains. Eventually we came to the site without me ever knowing the reason for the detour. Talk about bowing to the Japanese, I could have kissed them.

The job here was to commission and start-up a hydro-cracker, one of the most dangerous parts of an oil refinery which doesn't take kindly to mistakes, or take prisoners. The language barrier was a big problem as the Japanese hardly spoke any English. They had a Polish mechanical supervisor and then there were Italian, Russian, Japanese and local contractors on site – oh, and a Geordie. So communication was not exactly the strong point of the project, a worrying thought as Iran is hardly the place to be misunderstood. A strange thing I'd noticed after working on a few contracts like this is that sometimes I'd find myself shaking hands and speaking 'pidgin English' with my mates in my local back home. You seem to be always shaking hands in these countries and trying (and failing) to be understood.

I was, for want of a word, 'billeted' in a block of small rooms with some of the Japanese. My room was so small that when you got out of bed you bumped into the opposite wall. There was a small table, a wooden chair and a one-bar electric fire. This was up in the mountains of Iran, winter time and the mornings were so cold that you got out of bed, switched on the fire and then quickly got back in and waited for the room to warm up. The place was like Tenko. It was fenced off and had armed sentries up in watchtowers all around the place. I didn't quite find out if they were there to keep us in, or to keep out any undesirables, but I wasn't about to test them. In the mornings, we used to bus it into the plant from the camp and would be greeted in the office with a horrid pong! The Iranian tea boy (who was also the cleaner) would for some reason, just before you arrived, panic and start to mop the passageways with an old stinking mop, minus disinfectant or anything that had one's nose's approval. It stank like he'd dunked the mop down a toilet that had not been flushed for a week or two. Needless to say, I used to make my own tea and coffee, accompanied by memories of Baldrick making a 'cappuccino' for Captain Darling in Blackadder, the First World War series, which I won't dwell on.

The commissioning office was open-plan, which I personally do not like. The Japanese commissioning manager had the top desk, which was a tier higher than all his engineers who were positioned in front of him as if in a classroom. 'Sod that,' I thought and found my own desk behind a wall, out of the bowing and pecking order. As the coordinator, I had to attend meetings for all the planned work for that day. If a target had not been achieved, the departmental head in question would be shouted at and made to look a total failure, then stand there, head bowed, almost in tears. My turn came and I would immediately go into attack mode, criticising their lack of safety. Not that they took any notice but it was my attempt to try to break this hideous bullying regime. The way they treated their own staff made me cringe but there was this British guy who used to toady up to them all the time. Although he was an engineer himself, he'd carry the Japanese engineers' tools for them and call them 'boss'. I hated seeing him cow-towing like this; he must have been pretty desperate. What a way to hold a job down.

As I had a dual role, I had more than one boss which was a real pain in the arse. Apart from the commissioning manager, I had the 'dynamic duo'. One was a fat, soon-to-retire-or-die operations manager and the other was a young staff career engineer, straight from uni and the design office in Yokohama, who'd never been let loose before. I nicknamed him 'Tin Soldier' as he wore these awful gaiters and was so mechanical; I thought he was a robot. One day, Fatman and Tin Soldier were looking for me, shouting my name all over the place. We were about to start a bank of thirty-two fin-fans for the first time. All I could hear was: "Hopley-san, Hopley-san!" (San is 'Mr' in Japanese). Now, as I've mentioned in another chapters, fin-fans can be quite huge, ten foot in diameter, and these particular ones were to cool hydrogen.

"Hopley-san, Hopley-san! Come quickly, we start."
"Just hold your horses, I want take a look inside first."
"We already take a look inside, not necessary, we start, we start."
Each of these huge fans were enclosed within a sturdy cage for obvious reasons. I spotted a cage door open with a ladder halfway inside it!

"Cancel start, I need to inspect inside."

"Why you want look inside? It is OK!"

"Look here, sunshine, if we have an accident in this country and a local is killed or badly injured you may never see the land of the rising sun ever again and if you are lucky they will execute you."

"OK, we take a look."

I checked that the fans were electrically isolated and that the fuses were locked away, as I didn't want to end up chips. I climbed inside the first cage and couldn't believe what I saw. A whole roll of 1/4" copper tubing and twenty-four scaffolding elbows.

That was that. I climbed out absolutely furious.

"Test cancelled."

"Why test cancel?" said Tin Soldier.

"Are you from this planet? I want to inspect all of them before any rotate."

"Only one necessary, all rest is OK!"

"If you think I'm going ahead with commissioning these fans without an inside inspection, you can sod off." I stormed off to the main office before I took his life in my hands – as in, strangle the little shit! I was greeted by Fatman.

"Everything OK? Finish?"

"No OK."

I made myself a coffee before smelly mop could get to a mug. He was busy filling the sugar bowls with brown fingernails. There were no toilet brushes. Fatman and Tin Soldier were in a heated discussion with the commissioning manager, who looked as if he'd been told to commit hari-kiri. He came over.

"Why you stop tests?"

After I had countered to ten, I told him.

"OK, you check, then we start."

After I'd quietened down, I headed out. Close at heal were the dynamic duo.

"I need two labourers to assist me."
"OK no problem, I get."

I then went into each fan cage. Almost every one of them had debris inside. The complete list is as follows: one roll of copper tubing, twenty-four scaffold elbows, one I-beam iron girder one metre in length, which was just waiting to fall onto the fan blades, welding rods, tools, an assortment of nuts, bolts and washers, chain blocks, wire-lifting straps, bottles of paint thinners, chunks of wood, pieces of plywood the size of a mattress, and bottles of urine. These bone-idle local contractors were sleeping in these cages. I headed back to see the commissioning manager and placed the piece of iron girder on his desk. "You wanted to know why I cancelled the test today. That's one of the reasons"

He didn't even look up. It might as well have been a piece of cotton wool. The fans were cooling hydrogen gas and if there were any leaks and an ignition there would have been one hell of a bang, followed by a huge fireball. There was nothing else to say.

On the test runs, I was accompanied by the dynamic duo and an Iranian technician, who had to witness the tests and, most importantly, sign the acceptance certificates. I started each fan, one by one, and some were not as quiet as others. Each gearbox had its own characteristics. I knew the American gearboxes well and there was nothing wrong with them but the slight difference in sound was noticed by the Iranian technician, who shouted: "No good, No good!" I tried to explain to him that it wasn't a problem. "No sign, no sign." Shit. The Japanese wore these fancy soft-napper leather gloves and I knew that he wanted a pair. I shot back to the office, got him some and, hey presto, he signed for the equipment.

I wasn't getting much sleep as the Japanese engineers were partying nearly every night and got smashed out of their minds.

These parties went on until two or three am, bad enough at weekends but every night?! As we worked seven days per week, it drove you bonkers. I complained to them but they were too pissed to take any notice. Although it was a dry country and a dry camp, they managed to get booze off the Italians, who were allowed wine with their meals. I was never invited – I wonder why? But there's nothing worse than being sober and listening to a crowd of drunks. Eventually, I moved out to a bungalow which I shared with a fellow Brit, a monkey-hanger from Hartlepool, who turned out to be a great person and a real good laugh. He was the manager of one of the main contractors and we needed each other's company, being the only Brits on site at camp Tenko. The toolbox-carrier lived in town with some Japanese, probably as a cleaner and cook as well. Life at Tenko was rough. We'd finish work, bus straight to the camp restaurant and eat fish and rice and, for a change, rice and fish. I like fish but every day! And raw fish, sushi, too!

The camp did have a games room. The Japanese sat round tables, playing mah-jong, each with a cigarette dangling from their mouth. Not one of them would speak to you, or even acknowledge your presence. I went to this place twice the whole time that I was there. For me, the only thing I could do was to have a non-alcoholic beer at the bar. The barman was an Iranian and would ask incessant questions. There was also a TV but the only thing on Iranian TV were trashy American 'how John Wayne won the war' movies. Why the Japanese liked these, I never know; they probably had a good laugh. So there was not much else to do but read and listen to the good old BBC World Service, which I think helped to alleviate my homesickness. What used to amaze me about the Japanese was that at seven o'clock in the evening, when we were supposed to finish work, if the boss stayed, everybody stayed. They had nothing to do and would be falling asleep at their desks but remained. I called into the office one night at ten pm and they were still there, asleep.

One day, I was reversing out of the office car park and noticed a sly-looking bloke watching every move I made. God knows who he was: he was Iranian but didn't seem to have a job as such. He was there often and used to give me the creeps. For most of the time I was in Iran, I felt uneasy and under pressure all day long and, of course, you had the piss artistry in the evenings. It was like slow torture. The project was getting quite out of hand. There was such minimum safety on site and in and around the huge reciprocating hydrogen compressor that we were preparing to start and test, it was unbelievable. It was a freezing-cold December day, we were up in the mountains of Iran and there was snow on the ground. I was carrying out pre-start-up checks underneath the compressor, surrounded by a multinational team of contractors. The place was covered with lagging, tin sheets and mounds of fire-protection cement, falling from everywhere. These contractors were standing all over fragile governor-control linkages, there to shut down the turbine if it went into over-speed and prevent it disintegrating. Instrumentation was being damaged. In some areas, I was knee-deep in all this crap - it was a nightmare. A huge Russian technician opened a valve on a pipeline that was still purged with nitrogen and, in a second, fell over unconscious. He was very luckski, as N2 knocks out the air. The local contractors, unaware of the dangers, managed to get him out. What a day. One could have killed for a drink.

The following day, three Dutch engineers, who had just arrived, had a meeting with the Iranian refinery director and his cronies. The Iranians all turned up late in ones and twos, looking dishevelled. Every time one of them appeared, the Japanese commissioning manager would stand up, bow and go and fetch them a coffee. I could see Dutch engineers getting impatient. I could also sense trouble. Eric, the Dutch mechanical engineer, and the swarthy Iranian refinery director locked antlers immediately. The meeting kicked off with ridiculous demands from the Iranians. We'd had problems with the hydrogen compressor, problems that we were working on, which were now being resolved but our solutions had to be explained in detail to the Iranian delegation. As they were the clients, they had the upper hand but we just couldn't get through to them. Tempers got nasty and the Iranian director began banging his fist on the table with venom. Eric flew to the whiteboard, glaring at him and spitting out the problems we had and the solutions we were using. I was expecting the next sentence to be: "And if you don't like it, then fuck you, we're getting our clogs and going." After a few tense moments of eye twitching, to our surprise and relief, the director said: "Yes, we accept your modifications," then stood up, shook hands, laughed and left, followed by his un-merry men. "Jesus, that was close," said Yan, one of the engineers. We could all have done with a Geneva (strong Dutch gin).

The Dutch engineers were a great bunch, especially Eric who was the joker in the pack. There was a Polish engineer called Voitek, in the vendors office. Voitek was a little on the nervous side but had a good sense of humour, which was just as well, considering the jokes Eric and I subjected him to. On one occasion we drilled a hole in his desk, just in front of where his chair was. We fed in some small-bore plastic tubing, so that it ran along and under the other desks that were next to his, and fixed it into the hole. We then attached it to a small handpump at the other end of the office. Voitek came into the office, sat down and started to write his report. Eric started pumping away and a few seconds later Voitek shot out of his chair with a crotch full of water, as if he had peed himself. Another time we made an horrendous-looking spider. Eric hung a small hook on the ceiling above his desk, and fed string through it. In came Voitek, completely unawares. How we kept a straight face, I don't know. He clearly knew we were up to something as he kept his eyes on us but we pretended not to notice. After about ten minutes, he was back into his report. Eric slowly lowered the spider until it was level with Voitek's head.

He still hadn't noticed with writing his report. When he finally saw the spider in front of his face, he yelled and shot out of his chair; his chair went over and he spilled his coffee all over his report. Voitek used to leave his site helmet on his desk. "Not again, Eric, the poor guy is going to be a nervous wreck." In walked Voitek from the main office. He saw us, grinned and wagged his finger. He looked under his desk, above his desk - no tubing, or spider, no coffee cup stuck down with glue. But he was definitely on his guard, or so he thought. We made coffee and brought one over for him; he was suspicious and made me taste it first. I did so. "See? No problem." He checked all around him again. Ten minutes passed; he was now into his report. Eric pulled the string that we had attached to his safety helmet. The helmet shot across his desk like a demented rat. Voitek shot up like a demented Pole. He was well and truly Pole-axed! What bastards we were but what a great person he was.

We were now getting problems from everywhere and under a lot of pressure from the Iranians, who were leaning on the Japanese, who were leaning on us to get the plant started on time. Unfortunately, the Japanese I was working with had a tendency to panic and fold under pressure. On this particular day, I was walking down the main refinery road from the office, hoping to bump into the monkey-hanger, as I had a few questions to ask him. He was also under a lot of pressure, as the Iranians for some reason wouldn't let him send for his British team. As I mentioned before, commissioning and starting a hydro-cracker can be a very dangerous business and, when you don't have the experts on site, it didn't bear thinking about.
  "You OK?" I asked him.
  "Am I fuck. Those crazy bastards want me to get my start-up team from Eastern Bloc countries. We've tried using them before and they're a bunch of useless arseholes! I'm going to Tehran tonight, straight to the Swedish embassy."

"You're kidding me?"

"Like hell I am, this place is gonna go up in smoke."

"Oh, just great, you're going to leave me on my own with these nutters? How will you get to Tehran without transport?"

"I'll pay over the odds for a taxi, or walk even. It takes four hours by car."

"You'll freeze to death! And anyway, how will you get out of Tenko? Remember the guards up in the watchtowers all around the place. It'll be your luck to choose a night when they're awake!"

He stormed off. I couldn't quite catch what he said, but it sounded like: "Something, something, Scouse git." I laughed one of those 'laugh when you are scared' type of laughs. What I saw coming towards me cheered me up no end: a black Mercedes full of unshaven, collarless oil ministers and a TV crew. What sent a shiver down my spine, however, was the coach full of Ayatollah Khomeini lookalikes following it. I could feel the hatred aimed at me. At that point, I had never felt so intimidated in all my life, even after being bombed in Basra, Iraq and involved in three Gulf wars. I headed over to where we were going to test a high-speed centrifugal pump, which was driven by a steam turbine. The vendor representative, a Syrian guy, had had lots of trouble aligning the pump to the turbine. As the Iranians would not allow any more Westerners on site, the companies involved were sending anyone who could get a visa. Although he was a decent guy, he was a complete arse at his job. Vendor engineers usually know their product inside out but this person just didn't have a clue. My suspicions were unfortunately accurate as he confided with me that he was in fact a sales rep and not an engineer!

I approached the test site with trepidation. The area was pure bedlam. In order to heat the turbine casing to allow for thermal shock, I had introduced steam to the turbine to bring it up to operating temperatures. The steam exhausted through an open pipe above, which had no silencer and made the noise absolutely ear-piercingly unbearable! The decibel count must have fallen off the scale. I think a Richter scale would have been more appropriate as I could feel my whole body vibrating as I got nearer. I put on my ear protectors and climbed the platform to start the turbine. There must have been twenty Iranian student engineers around the machine and amongst them was the Syrian sales rep mouthing some bullshit. What amazed me was that he was oblivious to the situation that he'd put me in. I shot him a look of: 'What are you doing, you useless bone-idle little shit?' He had let the temperature drop too much to enable a start. This meant another hour at least. The noise was awesome and surrounded by nitwits asking daft questions. "Mister, mister!" The Iranian technicians were pointing at my ear protectors. They obviously wanted a pair. Honestly, you could be in deep thought, trying to solve a problem, needing to concentrate, and they'd ask you what your name was and where you came from.

"Mister, mister!" They kept pointing. I had to get away from this madhouse; it was too preposterous for words. I got into my car that I'd left by the turbine and headed towards the office for a coffee and a break to gather my thoughts. Then I caught a glimpse of myself in the rear mirror. What the...? I was covered in this thick black grease. Eric, the bastard. Before he had left, he had put Molykote grease on my ear protectors, knowing that I'd be using them the next day. That was why the crazy gang were pointing at them. Who's the jackass now? It was all over the place. I headed back to the house and into the shower. Not only is this type of grease lubricant black, it is also waterproof. It took ages to remove and was all over my clothes and car. The thoughts of revenge on the Cloggies entered my head at the time. When I told my housemate, he nearly died with laughter. And Voitek? Say no more. I would never live that down. For weeks after, I kept finding this bloody black grease everywhere.

Christmas came and it was just like an ordinary day. We were all working. Smelly-mop had greeted us with the fine fragrant aroma of a bouquet of roses from a freshly mopped floor. Must have been stood in dog shit overnight and dunked in a bucket of cat urine. I grabbed a coffee, made by myself, and headed for the morning meeting. For a brief moment, I looked out of the window at the snow-covered mountains. My thoughts were of home and what my family would be doing. A wave of sadness, tinged with guilt went through me. "Hopley-san!" Bump! Back to reality. I somehow managed to acquire a bottle of whisky, courtesy of the Italians, bless 'em. We had been joined by another Brit who was working for a Dutch company. Both of us managed to persuade my housemate to grin and bear it and stay. Which he did, to my relief. The Japanese didn't celebrate Christmas at all but at New Year's Eve, they seemed to be going for it. All afternoon they were holding a crazy ceremony, where they would roll some kind of dough into a ball, put it onto a wooden block, then somebody would swing this huge wooden mallet. How it missed people's heads, I do not know - I wasn't looking. When the dough was flattened, it would be filled with some kind of sweet substance. I never found out what it was. I thought that we'd be in for a treat.

Just before midnight, they all gathered outside around this huge gong. At the stroke of midnight, we took it in turn to bang the gong. This lasted for about five minutes, then they all evaporated and the three of us went gloomily back to the house. We had come up trumps with the Italians again but the only drink that we could get was a bottle of... er... it looked like vodka and had a mysterious plant inside. We sat there looking at it and said at the same time, "Shit, let's drink it." The only mixer we had was orange cordial but at least we did have an alcoholic drink, which was better than nothing. We ended up having a good laugh.

The big day came and time for starting the hydro-cracker. The Japanese engineers were nipping everywhere like scalded cats and if I heard my name called once, I heard it a thousand times! When you start up a refinery for the first time, especially a hydro-cracker, it can be daunting at the best of times and this was definitely not the best of times. You had the Japanese engineers shouting and running into each other and the Iranian clients. Imagine this: the plant had just been started, you are listening for all kinds of things - bangs, swishing, whole pipelines hammering, vibrations, smoke, fumes and alarms going off. You are also monitoring temperatures and pressures, making sure the whole thing doesn't blow you to kingdom come, when a pair of Iranians would come past holding hands. "Hello, mister, what is your name? Where are you from?" I'd have loved Salvador Dalí to paint this.

I called into the control room, which had huge steel doors and walls about twelve inches thick. 'Lucky them,' kind of went through my mind. My mate was looking at a bank of monitors. "How's tricks?" I asked. He nodded at the screens that were showing the temperature of the reactors rising, all of them. "We're going to get a runaway if the temperatures don't fall." Great. I had to go outside again. It was a ghostly scene, dark with hazy smoke. I could just pick out the Tin Soldier. Either he had balls or he wasn't aware of the dangers that surrounded him. "Mister, mister, what is your name?" came from out of the haze. It would have been funny had it not been so terrifying. I looked at him with his bad teeth and American flag on his helmet and thought, 'This must be as close to Hell as one can get.' The only difference being, I would assume, that there's only one devil down below, whereas this place was full of them.

The commissioning and start-up was successful and the place didn't go up in smoke. I said my goodbyes to my mate the monkey-hanger, got in the car, and who should I be sharing it with but the Tin Soldier. He was leaving that night too, although not on the same plane, thank God! We headed for the company guest house, with not a lot of conversation, only thoughts. It was 3pm and my flight wasn't until 2am the next day. The driver wouldn't be here until midnight. I watched some videos and managed to get some sleep. The drive to the airport was sinister, as there were armed police and soldiers all over the place, obviously some bigwig was coming through. At the sixth checkpoint, I was getting a wee bit concerned, as time was passing. I finally arrived at the airport, only to be confronted by more armed police. "Come, come, passport, passport," said the armed policeman, trying to make it as awkward and frightening for you as possible. Immigration was the next hurdle. I waited in the bustling queue, finally reached the counter and handed over my passport to the young official.
"English?"
"Yes."
"Where you been?"
"I have been to hell, sorry, working on an oil refinery in Arak." (A city in Iran.) He aggressively opened the passport with a sly smirk and logged onto his computer to check me out.
"You engineer?"
"Yes."
"Why you come to Iran?"
I had to stop myself from saying, "God knows."

I had to be careful not to speak any Arabic. They speak Farsi in Iran and Arabic in Iraq and of course they had been at war with one another. So to be on the safe side, I just spoke English as I didn't fancy being jailed as a spy. This was about the longest ten minutes I have spent. He threw my passport down on the counter. "OK, you go." I had to suppress my thoughts away from my tongue. More searches, through passport control... I climbed the stairway to board the plane and they were still checking passports and boarding cards. I was expecting the captain to check as well. We took off and a two-ton weight lifted from my shoulders. It was a Lufthansa flight but if Hitler had been serving drinks, I would have kissed him after kicking him in the balls. Old Douglas Hurd was spot on with his gangster state statement.

OIL RIGS
1988

(In which I nearly don't survive a survival course.)

Working for one of the main oil companies through an agency, one tends to get lumbered with the shit that their engineers don't want to do or can't be bothered to do! This time it was BP, based in their office in Dyce, Aberdeen. I like Aberdeen, the granite city, very much as it feels like a second home to me. The job that I was tasked to do was to gather information on all the new mechanical equipment on board a semi-submersible oil rig that was to be based in the North Sea and was currently undergoing a major overhaul and upgrade at a shipyard in Loch Kishorn, a sea lock in the north-west of Scotland in the Highlands. In winter time! I packed my coat. (I wonder if Aberdonians are aware that the granite in their building material gives off radiation? Just a thought!)

I choppered up there and was greeted by a whole lot of the white stuff. I don't know how the pilot could see in a blizzard but we did make it coz I am writing this. It was bloody freezing as it usually is North Scotland in winter but my accommodation and office was nice and cosy and the coffee was hot and free so no complaints so far. The job was not going to be a piece of cake as there was so much information to be gathered. I sat at my desk and reread my scope of work that BP had hired me to carry out. After a great deal of coffee, I decided to start with the big stuff, i.e. generators, pumps, compressors, valves and piping. The weekend came quite quickly and BP were throwing a big bash due to the completion of the refit/upgrade. It was to be held at a huge manor house by the loch. I received, I think, a belated invitation. I must have been an afterthought or somebody felt sorry for me! As it was January, the Jocks were still in a Hogmanay-ish mood. As I sidled over to the bar to get a wee drink, and some side glances were aimed at me as I ordered a drink with my Scouse accent. A Sassenach amongst the Scottish throng! Better not mention football, I thought, with me coming from Liverpool.

I got into conversation with someone I didn't know, nor was introduced to. The conversation was obviously about the rig, then moved on to world affairs and the price of oil (what else?), when I felt myself being propelled backwards with the help of my boss. When we were out of earshot, he informed me that the person I'd been talking to was a BP director. "What's the problem with that?" I asked, but he put his finger to his lips and gave me a nod to follow him. Then he disappeared. God knows what Mr Director thought about that. It seemed my creepy boss didn't want his boss to know what I, a mere agency engineer, thought of the mighty British Petroleum. Oil-company staff tend to behave like this, kowtowing to a hierarchy that they invent themselves. I got myself another drink. Years later, I encountered Creepy Boss working for an agency. Seems he'd had years in a nice cushy well-paid staff position with BP, had been made redundant with a fistful of dollars, then got a himself another cushy number. I would bet it was a quid-pro-quo basis, after him giving work to them whilst at BP. That's how I see it and how things work in this game!

I stayed in Kishorn for another month until the rig departed out into the North Sea, then it was back to Aberdeen in the BP office for several more weeks. I needed to go offshore to gather further information as most of the operating and maintenance manuals were now obsolete due to the upgrades that had taken place on the rig. I hadn't seen much of Creepy Boss since arriving back from Kishorn, thank the Lord. Having not been offshore before on a rig, I was a little apprehensive and excited but first I had to go on and pass a survival course as you are not allowed offshore without a valid survival certificate.

You've read the previous chapter on survival courses so you'll be prepared for some of this. A survival course basically involves being subjected to a scene from a disaster movie (think The Poseidon Adventure or Towering Inferno). I've told you about helicopter dunkings but believe me, this is not the worst thing on a survival course; the worst thing is when the firemen get their paws on you, give you a mask and put you in a smoke-filled room full of obstacles. I made a stupid mistake of putting my mask on and taking a deep breath instead of purging it first. Obviously, it was full of smoke. I managed to negotiate through all the many obstacles and sustained many bruises and learned a few more curses and expletives as well. When you've survived that, they put you into a succession of tubes, just wide enough for you to move. It's pitch black inside and God help you if you suffered from claustrophobia as when they close all the exits, the feeling is of sheer panic. One of the students did panic and burst through the tubing. (I heard that he was the Lord Mayor of Liverpool at the time so used to getting out of tight spots no doubt!) After this, you have to negotiate your way through a sewer pipe full of black smoke – a new unused one, we all hoped! This time, you aren't given a mask and must get out of there rather quickly before you suffocate as you cannot breathe nor see. When/if you do get out coughing and spluttering, you come to something like Dante's Inferno - a huge warehouse full of thick smoke and flames, definitely not a recruitment centre for the fire service.

One particular time, I teamed up with two French engineers from Total the French oil company. We had to locate a mannequin inside the smoke-filled tubes and bring it out. I was in front of them when I located it and shouted back to let them know, then suddenly found myself being pulled backwards instead of the bloody mannequin! Next up, inside sewer pipe, full of black acrid smoke and maskless, they both froze in front of me and blocked my exit. I gave the one in front a 'little pat' to get him to move his arse before we all asphyxiated. In yet another fire drill, we had to put out a helicopter fire and rescue the pilot. The ground around the chopper was ablaze, I was the one to rescue the pilot and they were supposed to douse the flames with water from a fire hose but instead of dousing the flames, they directed them towards and around me. Talk about misunderstandings. I was now in the merde, excuse my French. Sacré bleu! On my way back to my hotel I suddenly developed a thirst. The first two pints of lager didn't touch the sides. I was covered in soot and smelt like a bonfire. I didn't half get some strange looks at the bar!

Anyway, back to the job at hand... At the helicopter check-in at the heliport, you give your details in and are given a survival suit with a number on which I assumed had been modelled by the Hulk. I'm approx 5ft 9ish so I pitied the little'uns on our flight. The suits are very uncomfortable as the neck nearly chokes you. It's to stop the ingress of seawater but the necks, unlike the rest of the suits, are modelled by an actual elf and are so tight around your neck it's difficult to swallow. No doubt they are also used as body bags when and if you are recovered from the sea after a ditch! It would help the rescuers to identify you.

When you are flying over the North Sea, which is rough most of the time, you think to yourself: 'Jesus, if this does ditch, survival suit or not, I possibly wouldn't survive three minutes.' Actually, as I was told at the survival school, you could/may last in the sea due to hypothermia, survival suit or not. Of course that's if you managed to get out of the now-upturned and sinking chopper with freezing salty water in your eyes and get past the struggling North Sea Tigers (offshore oil men) nursing their last-night onshore hangovers.

On arrival at the rig, you go to the helicopter reception and are allocated a room and a lifeboat number. My first such room had six double beds in it, toilets and showers, so it wasn't the Ritz but it was convenient and clean - in other words: a contractor's room. I introduced myself to the maintenance supervisor, had lunch and made a start in the afternoon after first acquiring a permit to work: God's unwritten law on an oil rig. The rumour spreaders' conversation at the dinner table was of the sister rig of the one I was now on. It had 'turned turtle' with loss of life. This knowledge, coupled with the storm outside, didn't help one sleep. You always get professional rumour spreaders on oil rigs (and everywhere else). Then of course I heard that it wasn't a rumour. So it kind of put the dampeners on my first trip offshore.

This type of oil rig can operate in shallow and deep waters. It can be towed from well to well and can be positioned immediately over the main oil-well valve which is called a 'Christmas Tree'. This is then deballasted like a submarine: the huge pontoons on each leg of the rig are filled with seawater to the exact level until it sinks to the required depth and the final positioning is done by the thrusters on each leg of the rig, then guided by dynamic positioning (run by computers) over the main oil-well valve - the Christmas Tree, so called due to the fact that it looks, well, it doesn't really look like a Christmas tree, it's just that it has all sorts of items sticking out of it: valves, pipes, connections and branches. Then when the rig needs to move to another well, it uncouples, jettisons the seawater and gets towed to another well and so on. There you go, you are now on your way to becoming a North Sea Tiger.

I left the worst job til last as I was feeling rather apprehensive about it. What it entailed was going down each leg of the rig, which was a good way under the water as the rig was now over the well and had been lowered on the main oil valve which was on the seabed. So it meant going along the connecting channels to each pontoon, which are interconnected, to gather my information while thinking that just an inch or so on the outside of the pontoons was the sea. I used to have these same thoughts when I was a sea-going marine engineer. I'd touch the bulkhead and just think that a mere inch on the outside was the Atlantic Ocean (as of course a ship's engine room is below sea level). Here I was again, underwater in the middle of the night and it was such a weird feeling. I gathered the required information and got out as quickly as possible! I have experienced many strange and dangerous things in my job around the world but this was not one to dwell on.

So it was back to BP Aberdeen to complete the job. My next venture on an oil rig was again set in Aberdeen. I had an interview with a major offshore contractor. The position was as maintenance supervisor on a Shell oil rig in the North Sea, way up north in Norwegian waters. The interview was successful and I got the job. First, I spent a week in the office to read up on the rig and the project. The company had secured the contract to maintain all the Shell oil rigs in the North Sea, which was huge for them. On week two, I went on the obligatory offshore survival course which incorporated a course of writing work permits as I would be responsible for procuring and issuing them to the technicians and engineers on my watch. We had to complete these massive forms before we left the college. I was in a panic as I had a plane to catch for a whistle-stop visit to my home and I just made it. (I was told that the permit forms are kept by the oil company's for ten years due to the Piper Alpha tragedy.)

The big day came and I was the first of our company to go to the rig and must admit to being again somewhat apprehensive going on my own but somebody had to do it. On my arrival at the helicopter reception, I was greeted by the mechanical supervisor and taken to the supervisor's office. He was ahead of me and as he got inside I overheard him say to the other people in the office: "They're here." There were two other Shell supervisors sitting at their desks. Basically I was there to take over all their jobs, so can you imagine what I felt like? Like sitting in a bath of freezing water. There were three supervisors: electrical, instrumentation and mechanical. I now had all their departments to run! And had to share the office with them all.

The next full day, all my crew arrived. The Shell supervisors sat back and said more or less: "It's all yours now, mate." I now had all the responsibility of the maintenance and engineering and crew of this huge oil rig in the North Sea. To be perfectly honest, I was scared. The thought of it made me feel sick. I have been through these times in my job before but I took a deep breath and carried on… growing an ulcer! I had arranged for the crew to see me in the conference room to introduce myself and get to know them and tell them about my training, qualifications and experience. So that was the first morning; the afternoon was spent with the crew showing me around the rig and the jobs that they were doing. Then I spent the remainder of the day going through the outstanding work, prioritising it, then writing out all the work orders and permits for the next day. You can imagine the help I received from the now-defunct Shell supervisors. I admit I felt sorry for them as if I was in their shoes – maybe I would react the same, who knows? Their company were making them redundant and I suppose they were waiting/wishing for me to make a mistake and sink. My technicians would come into the office with a problem and go first to the Shell supervisor that they were used to for their discipline, who would then direct them to me and sit back and watch. This was perhaps a smug moment for the three of them but I don't blame them.

The mornings started with the crew having to pick up their work permits. I was also getting it in the neck from the permit staff. If you hadn't dotted the i's, or crossed the t's, or had made a spelling mistake, they wouldn't issue them to my crew who were unable to start work, so I was getting agro all round. My first trip was like this nearly every day but I pulled it off without sinking. Soon my back-to-back came on board and I stayed with him for two days to bring him up to speed, planning and plotting before I went on leave. My first impression of him was that he was a bit 'bolshie'. He was a different discipline to me, him being instrumentation, me mechanical. 'Never the twain shall meet' came to mind but you just go with the flow.

Leave comes, you have written your handover report, rig safety report, update of the work being carried out and what permits are live, and you are waiting for your back-to-back just below the helipad. The chopper arrives, it lands, the noise is already a crescendo and you try to converse, you then head to the chopper. Even though the rotating blades are higher than your head you still have the tendency to duck. Ordeal nearly over, you climb on board, feeling for the seat belt and noting where the escape exits are located and the release mechanisms. As you are flying along back to the Shetlands you feel a huge weight has lifted off your shoulders, you then board an old turbo-prop flight to Aberdeen. You'd sometimes hear a loud bang on the side of the plane, which was a large chunk of ice off the wing being blown by the propellors that hits the side of the plane but you don't know that at the time and start to think: 'Have I missed something out?!' It had been quite an experience.

On my next trip, I found the desk had been repositioned. I like to be facing the door, old habits, and with my back to the three Shell supervisors. Slight changes had been made to some of my ideas without first discussing it with me which I thought was out of order and a little arrogant – there are two of us but you just had to go with it! It didn't take long to move the desk the way I preferred it to be. It may sound strange and or petty but it all helps you get through the day and made you feel at home. I was just beginning to get to know the crew, which was just as well because as I was then informed by the company, at this early stage, to carry out an evaluation/assessment exercise on all of them. I couldn't believe it as I hadn't had long enough with them but I had to do it. I interviewed each one in the conference room, away from the prying ears of the enemy, i.e. Shell staff. At least this was now going to be a once-a-year exercise, thank goodness. To be honest, I could have done with the help of a psychologist as their personalities varied quite considerably but it helped with my next task which was to select a leading hand for each discipline. With me not being quite used to them yet, there were some that disagreed with my assessments and choice of lead hands, which was fair comment. I think I chose well with the little time that I had. I sifted out the knowledgable, the resolute, the weak-willed and the awkward types; of course I got it in the neck from certain jealous and resentful elements, mainly the older technicians as I had chosen younger ones, sticking my neck out somewhat but it was a good exercise that I didn't enjoy.

So my crew were now up and running like clockwork, and it was now up to me to tweak and make it Swiss clockwork! I had some doubt with the youngest of the new lead hands as, first thing one morning, I had a young female cleaner in my office making a tearful complaint about him. She was saying he'd made a derogatory remark about her. This was a serious and dismissible offence and was a first for me. Of course the culprit didn't want to be fired and the victim wanted retribution but I didn't want to lose him. I hadn't been trained for this situation but it had to be dealt with as all the personnel on the rig got to know about it. So I got him in my office and told him to buy her a big box of chocolates from the rig shop, write her a letter of apology and apologise to her face. Then I put him on probation to give him a chance to pull his socks up otherwise his position as lead hand would be untenable and would go to another technician.

The days were flying by and I was enjoying the work as stressful as it was, funnily enough, when 'boom' another problem hit the fan. It was morning time again when the O.I.M. (offshore installation manager, who is like a captain of a ship) called me into his office in a very angry mood. 'Shit! What the hell is the problem now?' He was like an explosive artillery. I lost my temper with him and said: "Hold your horses, I have just come into my office, please let me know what the problem is and I will try and resolve it." This quietened him a little, although not a lot. Apparently, during the night, while I was asleep and oblivious, the rig had shut down and lost squillians of £'s. This is an O.I.M's nightmare! And was now mine! I said I'd investigate and get back to him. It turned out that the instrument technicians had been making adjustments to some of their equipment and ballsed it up, so consequently the rig shut down and no one had told me. I spoke to the technicians involved but the O.I.M. wasn't satisfied with that and I had to write a report on the issue. So all my day's planned work was ruined and I had to start again.

The next thing I knew, Shell set up a court of inquiry in the conference room over the issue. I really couldn't believe this but it's a fact! They set it up like a courtroom sitting, where the O.I.M., the rig safety officer, two of Shell's ex-supervisors (who I'm sure were enjoying the theatre), oh and yours truly. Each one of the technicians involved had to sit facing us and tell their version of events. All this was very much due to the tragic events of the Piper Alpha and the Cullen Reports. Nobody was found to be individually at fault but we all received a caution and reprimand from the O.I.M. So feelings were (and still are) very high on oil rigs in the North Sea. The Kangaroo court ended with poignant and misgiving feelings all round. Then it was back to bringing up more of the black stuff!

One thing I recall still sends shivers down my spine. I was flying out to the rig, again in wintertime, with very dark skies and gale-force winds and lashing rain. The chopper had to drop off some men at another rig before me, then we began to take off to go to my rig. The chopper lifted about six foot then aborted take-off. The pilot attempted three times as the high winds kept it down and each time you could see the white tips of huge dark waves over the side. A huge Scotch would have come handy at that moment. These are just some of the dangers that personnel who work on oil rigs go through on top of the work. I was alarmed to say the least, I don't mind telling you!

The saying 'chalk and cheese' springs to mind on my next offshore work on an oil rig. This was in West Africa. Can you imagine the local workers on the drill floor wearing flip-flops? No, I couldn't either until I paid a visit to the company's rig offshore. The day started with the crew waiting for transport to the heliport. I was the superintendent for the maintenance of the rig but based in the office in town. It certainly opened my eyes after the stringent safety measures with oil companies in the North Sea. For a start there was no permit-to-work system in place, plus the oil company did not envisage starting one. The company was full of Italians and I doubted very much that they knew anything about the oil industry, let alone permit-to-work systems - more the laundry business (see previous chapter on Benin).

It was very difficult making sure your crew were safe and weren't going to have any accidents. This was an incident and a nightmare waiting to happen. I used to pay a visit offshore to the rig once a week. The heliport check-in was a waiting room full of local rig workers, mainly roughnecks, labourers and just about every scallywag in town. It stank of B.O. and other odours and was full of mosquitoes - those lovable little vampires. So not a very pleasant place to be and no chance of a coffee or a bacon sarnie. There were no survival kits or even a life jacket but here it wouldn't be the cold that got you in these waters if the chopper went down but those big fish with big mouths and big teeth. It was a big scramble to get to the chopper even though you were booked on it, but the locals just did not adhere to this and just about every other thing that should have been adhered to.

Then you get on board and find personnel walking about the rig wearing these bloody flip-flops and not wearing hard hats. I pulled the crew together for safety talks. You are talking to men that have been around the world on many jobs and it was easy to feel very foolish and inadequate talking about safety. You could only imagine them thinking: 'What a tosser,' but it had to be done. 'Teaching grandma to suck eggs' came to mind but you are the boss and bosses have to do these things. I had an instance where an American electrical engineer went offshore without telling me or anybody! We received an emergency call from his wife in America and I drove down to the quay as he was nowhere else. It was now dark and the last helicopter had now landed back at its base and nobody knew where he was or had seen him. A launch came in, tied up at the quay and lo and behold, he got off. I tore a strip off him. He wasn't worried about his wife and the emergency at home but wanted to fight with me! I was annoyed, frustrated and angry but took him back to base to telephone his wife then make arrangements for him to fly home ASAP and not to return. They say it's tough at the top. I hadn't reached that area as yet and I was finding it tough enough at my humble stage of the pecking order.

Angola was my next venture to the rigs. The helicopter check-in was in a huge corrugated aeroplane hanger just off the jungle. It was crowded and hot as hell where I waited for about two hours for the chopper to the rig. This was for Sonangol the Angolan oil company and even though it was a huge oil company, for at least this part of the world, there was still no safety in place. It was tread very easily again. Well, the first thing to hit me on the rig was the unmistakable pungent smell of H2S (hydrogen sulphide) meaning possible instant death! I got hold of a monitor and the reading was sky-high. We were supposed to work in this and there were leaks all over the place! I visited all their satellite rigs and reported lots of corrosion and leaks of just about everything and of substances that could be leaked. After that, I made sure I conducted business from my shore-based office but still made weekly visits and kept reminding my team to be extra careful and carry monitors all the time. I didn't add: 'switched on!' They all knew the score and fortunately none got a strong sniff of H2S. Again it was sucking eggs but I made sure I didn't get any yolk on my face!

You either like working on oil rigs or you don't. I think I was both – I liked the North Sea but not West Africa. You had to be alert and ahead of the game all of the time to stay alive!

Me at Loch Kishorn (a sea loch in the North West Highlands of Scotland) about to board the chopper to the North Sea rig.

Russia
1989

(In which I party in the dark and get my finger broken by the Mafia.)

"Take a blue company bag so you'll be easily recognised," was the company's advice. I boarded a British Airways flight bound for Moscow, wearing a black raincoat and carrying a blue bag. This was like a Freddy Forsyth novel. Oh, it got me recognised all right, by the wrong people. I was besieged, I think, by every illegal money-changer at Sheremetyevo, Moscow's main airport. Eventually I was spotted and rescued by the Russian interpreter. We had to go to the internal airport to catch a flight to our final destination, which was an industrial town approximately 300 miles from Moscow, called Nizhnekamsk, in Tartarstan. I had to wait until a deal was struck with a taxi driver who was going to be the lucky guy (which turned out to be unlucky for him).

We set off and hadn't been going long when it started to snow and it was getting dark. I was in the back of the taxi with a young gorgeous female interpreter and had a festering sourpuss of a taxi driver up front, who was now beginning to realise that his deal was obviously too low in roubles for the journey. I reached for the blue bag and brought out two small bottles of champagne, courtesy of the wonderful British Airways cabin staff.

"Bet you have never had champagne in the back of a taxi before?" I'm one smooth Scouser. But before she could answer, BANG! What the hell was that? We had a puncture. Sourpuss got out, cursed, as you do, and kicked the wheel. The snow was getting heavier and we had only driven about two miles. The wheel was changed and we set off again. Another mile or so and... BANG! Another puncture. SP got out again and I heard my first Russian swear words. I could have sworn he said, "Fuckingski bastardski." The moment was not quite right to give him an elocution lesson. This was unbelievable. How can we possibly get two punctures one after another in the space of two miles? One is a pain, two is damn unlucky and over the top in unlucky stakes.

"Mafia."

"What, Mafia as in concrete shoeboxes and Godfathers?"

"Yes, they wanted to change the money with you at airport."

"So they sabotaged the taxi?"

"Welcome to Russia."

SP had to flag down a vehicle to buy a wheel so that we could continue our journey. The international airport in Moscow had been dark and unwelcoming but this was bleak; there was nowhere open and it was only eight o'clock in the evening. It was cold and with only wooden chairs to sit on, this was going to be a long night. A coffee shop opened about six in the morning. The coffee? Well, it was hot. Hard-boiled eggs and stale bread was on the menu. It got to about four in the afternoon and there hadn't been any announcements at all. Then all of a sudden, a gate opened and this huge woman shouted "Nizhnekamsk!" in a rather deep voice. Bodies appeared from black holes and wooden chairs. The mad scramble was on. I managed to somehow get through and approached the aircraft, which I did not recognise. Then I felt an overwhelming feeling of doom as I saw the word 'Aeroflot', known by many a traveller as 'Aeroflop'. As I climbed up the steps, I noticed that the captain and co-pilot standing at the top. When they saw me, an obvious Westerner, the captain moved his foot to cover a hole in the carpet. The smell inside was like a farmyard. I was quickly ushered to the front, to a seat minus armrests, by a burly stewardess. I could see another plane thundering down the runway, or I should say 'sloshing' down the runway in the snow and mud. We started to accelerate and up we went. The aircraft was now at an angle of 45 degrees, when the two rotund stewardesses, 'Boris and Doris', unbuckled their seat belts, stood up, yelled and ran to the rear of the plane. I honestly thought that my number was up and that we were going to crash. I looked round expecting to see a fire or something but to my great relief it was two drunken farmers having a fist fight. Boris and Doris chinned them at 45 degrees. They slept most of the way. Later into the flight, I was served a sweetish drink out of a brown melamine cup and some dry biscuits. I was not looking forward to the next three hours.

When we landed, we had to wait ages to disembark. Two policemen the size of outside toilets came on board. They were wearing huge grey overcoats, jackboots and ushanka (Russian hats). They asked the farmers, in their polite Russian police way, if they had "a nice enjoyable flight?" I have never seen drunks sober up so quickly in my life. To say 'frogmarched' would not have been an exaggerated statement. As I walked down the gangway, sorry 'steps' - my merchant navy days creeping in - another rotund woman in uniform bawled out: "Tourist, tourist?"

"Business," I replied. She just kept repeating: "Tourist, tourist!" until it dawned on me that she must be some kind of official who orders foreigners about. To disobey would have meant certain death, or at least a vacation at the nearest Gulag hospitality centre. I was escorted to a gate in the perimeter fence and not through the airport, which I thought most odd. Outside was a coach with only the driver on board, who started to beckon me inside like a lunatic. I hesitated for a while, then I saw the ubiquitous stickers: 'Texas, the big country', 'Don't mess with Texas', 'Drink Bud', 'Boro for the cup' and various oil company stickers. I had arrived and this was obviously my transport.

After a disappointing and scene-less drive through a baron industrial area, we eventually arrived at the apartments, which were dismal. The company had rooms in two apartment blocks right next to each other. They had no names only numbers, Clint Eastwood style. The driver was very impressed with the apartment and kept making gyrating movements with his arms as if he was showing me the crown jewels.

"Very good?"

"Oh yes, very good."

The walls had been painted with a distemper-based paint. I remember it as a child, the bloody stuff used to get everywhere. Why we used it I don't know but it certainly had the right name 'distemper' as it got my temper going. When you had been out for a drink, you were sometimes inclined to recline on the walls, and the next day you'd be covered in this bloody awful cretaceous chalk. My black raincoat soon became pale-green. No wonder dogs go crazy with distemper (not the same thing but same spelling).

The rooms had the ambience of a junk shop. The bathroom had this huge bath and I'm sure that if you slid down while taking a bath, it would take you about two minutes to reach the bottom and you would need a decompression chamber if, and when, you surfaced. There were crude holes in the walls where the hot and cold water pipes came through; the joints had been welded and the scorch marks from the welding torch still remained. No attempt had been made to paint or chalk the walls. But I must admit that the water, after the rust had passed through, was red-hot. Being Russia it would be red! The radiators in all the rooms were like the ones from my childhood days, huge cast-iron type that were in the schools in the 50s. The drawback was, they didn't have any valves, so you couldn't regulate the temperature, which was at times stifling. In the mornings you would be looking out of the window, waiting for the coach to arrive to take you to work and you would roast in your thermal underwear, shirt, sweater, body warmer, coat, heavy woollen socks and high boots, as you wouldn't have time to put them all on when the coach arrived. By the time I left the apartment, steam would be coming out of my ears.

The kitchen was like something out of 'The Flintstones' – no, that's not correct, the Flintstones had a modern kitchen. There was huge cooker that had a gas bottle piped to it and when you turned on the gas, you could have roasted a whole cow in about fifteen seconds flat. On my first night, I had been invited to dinner by the company accountant and his wife, a German couple who I think had taken pity on me and, as I had only had a boiled egg and a piece of stale bread in twenty-four hours, I was famished. After dinner and some polite conversation, I thanked them and headed to the company bar, which was appropriately named The Back-Stabbers Arms, as it eventually turned out to be. Funny, no matter where you go in the world, there are always back-stabbers, plenty of them. You could bet your house on the conversation, as you will always get a "when I was" story. "When I was in such and such a country…" How they fooled customs by bringing beer-making kits through certain countries etc. I remember doing it myself. (I'm doing it now!) The things that you did when you were younger…

Talking about back-stabbers, the ultimate back-stabber for me was an Irish guy called Bennet, known to all and sundry as 'Bullshit Bennet', with whom I had had the displeasure of meeting many years ago whilst working on ships in Liverpool. This is what I call my 'Bennet problem': he was a loud and crude Irishman, full of the Irish blarney but the difference was, he was a malicious bastard! On one occasion, there were just him and me in the ship's engine-room workshop, when he extracted methane from his bottom. I said, "You disgusting fatherless person," but instead of laughing it off, he got all nasty and denied that it was him. As there were only the two of us in the workshop, I wasn't postulating and found it quite hard to believe that he denied my accusation. And that was only a bloody fart.

As Bullshit didn't have a car and was also my boss who, unfortunately for me, lived close by, I used to give him a lift when we were working on the same ship. One morning, I had overslept and was late, you know what you are like at twenty-two years of age and just married. The ship was, as we say in Liverpool, 'over the water,' meaning that it was in Birkenhead, across the River Mersey. He had to make his own way there by train and bus. I happened to be working on the ship's main engine and even though both diesel generators were running, I could still hear him on the deck cursing, swearing and criticising me in a very nasty way. I had saved him time, effort and money by giving him lifts to and from various ships and I had overslept just this once. When he saw me, I got all the verbal diarrhoea that his temper could muster. I told him politely to 'go forth and multiply' as you do. He made my life a misery after that and would stab me in the back at every given opportunity. I've found on my travels, there is always, always a 'Bennet'. Frightening, isn't it?

Anyway, I walked into the Back Stabbers and was greeted by 'who the fuck is he?' stares and a couple of floppy handshakes as if I had a turd in my offered hand. At this early stage in the evening, it wasn't hard to pick out the bar supporters and the bar supports, the red-nose brigade. One was a grizzly Irishman (nothing against the Irish, just that they seemed to be everywhere). He sported a hillbilly beard, a bit like Catweazle. I was later to find out that he used to get so pissed, he had to spend the next morning walking around the plant in sub-zero temperatures to sober up. Another was a lunatic piping engineer from the Lake District who hated everyone that didn't share his liking for the Rogues, sorry, Pogues. The construction manager, who was serving behind the bar at the time, greeted me with, "Welcome on board, I see that I have chosen the right man for the job." As I had never met him, nor had any conversation with him before, I wondered how he could have assumed this at such an early stage. Not that I was going to disagree with his assumptions. As the night wore on, I had to suffer constant torment from The Pogues, followed by wild shrieks from their greatest fan, the nutter from the Lake District. He glared at me, I put my thumb up and mouthed "Great music," just to stop him from smashing a chair or bottle over my head.

The next day was Sunday, so I decided to go for a walk to shake off an awful 'Pogue-over.' I had heard at the Back Stabbers somebody mentioning a go-kart meeting and I made some discreet enquiries so that I could get away from the crazy drunken throng and headed there. As I walked along, I had this constant feeling that somebody was following me. I was told later that it was 'internal security'. On another occasion, I was heading back to the apartment and sensed that I was again being followed. I stopped and pretended to tie my shoelace. When they caught up, I asked them why they were following me. And was informed in perfect English: "For your protection."

"From whom?" "From the Mafia." Say no more. I started to like the internal security, whoever they were. I thought it better not to ask!

On my first day of work, we arrived at the plant which looked old and clapped-out. I was introduced to all of my two staff, a drunken Jock who had the DT's every day, and an old fart, who was only a bit younger than Leo Tolstoy. I was sat at my desk when a delegation of Russian engineers marched in, followed by an interpreter. A technical drawing, all in Russian, was unfolded before me. The engineers would point to something and speak to the interpreter, who would then translate what the problem was to me. I would then try to solve it. The problem-solving would start by putting it into their 'modus operandi' which went: 'The seller advises the buyer to take such and such a course of action, to resolve whatever the problem was.' I would then have to get it vetted by my boss, who spoke fluent Russian, then it would go to the interpreter's office for translation. You would not know if they took your advice or not.

On one occasion, two strange Russians came into my office, and said: " Come, come," and beckoned me outside to a waiting car. 'What is this?' kind of crossed my mind, as did 'Where are they taking me?' as I have read Aleksandr Solzhenitsyn's 'The Gulag Archipelago' and it did make me feel like visiting a toilet. After about an hour's drive, without any communication, we arrived at another plant. They wanted me to witness the testing and signing-off of relief valves, which was a great relief. Relief/safety valves are used on oil refineries, gas plants, oil rigs etc. It is like a fuse, a safety device: if too much pressure builds up inside, say, a vessel, boiler or an oil tank, the relief valve will open and vent the pressurised vapours to atmosphere instead of the vessel blowing up. As you can imagine, relief valves are somewhat critical and if they are not calibrated and tested correctly they can cause all sorts of nasty problems.

A short 'When I Was...' story: one time in Iraq, I was standing on top of a huge live steam boiler. The temperature was about 100 degrees and I was working alongside a Lloyd's surveyor adjusting one of these valves. The surveyor would signal to me when the pressure of the boiler dropped. I would then signal to an operator to close the discharge valve to build up the pressure inside the boiler. When the valve opened, we would know the pressure that it vented off at and adjust to suit. It was, to me, always an unpleasant experience and a job that I loathed doing. You would be on top of a potential bomb, wearing ear-protectors. You would know that when the valve opened, it would make an enormous roar but it always made me jump. You can feel the pressure building up, like when you blow up a balloon, and when it blew, boy, you knew. It was always a good idea to carry a toilet roll with you for this job. How did they manage before? I thought of Chernobyl and shuddered.

Nizhnekamsk was a purpose-built town for people working in the oil industry. There were row after row of dismal grey blocks of flats, no street names (just numbers), and two restaurants, one called the Crystal. It could be quite embarrassing when you arrived at the Crystal, because usually there would be people queuing for tables and a rotund woman – it was always a rotund woman, I think she was the bouncer. She would see you and come running to put you in front of the queue. This was embarrassing enough but she would hit out at the people in the queue with a stick to clear the way for you. It didn't leave you with much scope for conversation. "I say, how are you? I'm British, pleased to meet you," so to speak. When she got you in the cloakroom, the real purpose of her kindly gestures was only too apparent: she would say, "Razor blades, bring me." They were probably for her, so my stock of blades was gradually depleting.

On my birthday, I hired the Crystal for a party. It cost me about £30 and we provided the drinks. I had invited all the expatriates, about twenty in all, plus some of the wives that were living there. The remainder of the guests were the Russian translators, engineers and their wives. We had got to know some of the internal security, or I should say: 'They had got to know us.' Two of their operatives, if that's the name for them, were doormen, yes, doormen. Let's call them Sacha and Sergey. During the course of the evening, we would send them bottles of Guinness and Benson & Hedges cigarettes. They told me that they were posted there to keep out any 'undesirables'. As the night wore on, I noticed five huge undesirables walk in unchallenged and sit opposite us at the other side of the room, which was a small dance floor. It didn't take an Einstein to work out who they were. The local Mafia had come to pay us a visit. They were more or less controlling many parts of Russia and certainly controlling this town. Sacha and Sergey had no chance of saying, "Sorry, chaps, you can't come inski," or words to that effect. They would have been spoken to, followed by a broken bottle shoved up their nose, or elsewhere.

I picked up a bottle of Johnnie Walker Black Label whisky and apprehensively walked over to give it to the obvious boss. I handed him the bottle and said something like, "Hi, you guys, and welcome, it's my birthday, please have a drink on me." They all started clapping and singing, "Happy birthday to you, happy birthday to you." They had baked me a cake, I blew out the candles and made a wish. No, I just made that up as you may have guessed. I shook hands and returned to our table wanting to cry as my hand was almost broken. I have a firm handshake but you could hear my bones crunch in their hands. I have seen some hard bastards during my time on the Liverpool waterfront. Plus, I have been around the world a wee bit. The Mafia boss was like Mount Rushmore, only harder! His minders – well, I would not kick sand in their faces on the beach – were even bigger. About a minute or two later, the mob boss came over to me with a bottle. "Have a drink on me." It was a bottle of one of Russia's finest pepper vodkas. "Er, thanks very much," and quickly added: "My friend."

This strange woman kept appearing from behind a ceiling support on the dance floor. I found out through one of the interpreters that she was a local dentist who wanted to get married to a Westerner and bugger off from the land of the 'frozen dentures'. After experiencing Nizhnekamsk, I could see why. A week later, I had an invitation to a local wedding at the Crystal and who should be there, but my infamous new friend, the local mobster boss. He walked over, sat down next to me, plonked a bottle of vodka on the table, then ordered a waiter to bring two glasses. I thought, 'Oh shit!' He filled the glasses and said "Nazdarovya" (cheers), knocked it back in one go and dared me to follow. I did. He repeated this several times. I thought: 'God, I am going to die.' I had to stop this madness. I pulled out a packet of cigarettes, as I used to smoke back then. One devil's pastime to another. I got out my lighter to light his cigarette. "What is this?" I had just swapped lighters with the groom. He studied it, said "Russian shit" and smashed it on the floor. "This is better." He pulled out a Ronson. The madness continued until I felt the room spinning and wanted to puke. I made some feeble excuse and left. On the way out, I fell arse over tip down the steps and broke my little finger, left hand. This I didn't discover until the next day, as you can imagine. You could have your head chopped off and not feel it after drinking that evil drink, with the devil himself. They actually sold vodka glasses with the devil on a swing inside them, which I forgot to bring home. My finger was now very swollen and painful and I had to go to the local hospital. The doctor looked at my finger and said something to the interpreter, who told me that he was going to inject a 'local anaesthetic' into my finger to enable them to straighten it. I looked at the needle and said, "No way." It was one of the old reusable types, about a foot long. Dr Frankenstein glared at me.

"Why not?"
"Because it may not be sterile!"
"Don't worry, we have boiled it."

I was about to protest, when: 'ouch,' it was in my finger. He was glaring at me as he pumped the fluid in. I glared back to make out that it didn't hurt. The next day I found that I could wiggle my little finger inside the cast, so I had to return. They made sure that they fixed it and fixed it they did. They put another cast on that went half way up my arm, which was a bloody nuisance. My little finger is still bent to this day. It's my 'Mafia injury'.

The trolley buses used to frighten the life out of me. They would approach virtually sideways, then the driver would hit the brakes and the bus would slew round and straighten out for you to board. There would be about half the town on the bus, so rather a tight squeeze. To pay, you would give your money to the next person, who would then pass it to their next person and so on and about ten minutes later your ticket would eventually arrive, via the same route in reverse. Most of the shops were very drab and others utterly depressing but I did manage to pick up some good bargains. For instance, binoculars, monoculars and telescopes were extremely cheap and were well-made. I bought myself a huge overcoat for twenty pounds that had been made in Gorky. The problem was, it has to be really cold to wear. I tried wearing it back home and got so hot that I wanted to kill someone.

Although Nizhnekamsk is not a big place, I still got hopelessly lost while out walking and there weren't any taxis at all. I asked some people who could speak English for directions, as my Russian was less than basic. They happened to be schoolteachers, who took the trouble to direct me to my Gulag Apartelago. Every time you went out and came in, the receptionist would look at her watch and record your movements, er, put another way, your comings and goings. You would end up calling out the time for her. I thanked the teachers and said goodbye. During our conversation, they had invited me to their school the following Saturday. I of course accepted. Saturday came and I had forgotten to set my alarm and was recovering from the night before at the Back Stabbers. The receptionist banged on my door and said: "Come, there are some school children wait for you."

'Oh no.' I quickly showered and dressed and was escorted to the school by these eager kids. The school was very basic and featureless and, again, no name, just School 42. How awful. I couldn't help but think of Douglas Adams' 'The Hitchhiker's Guide to the Galaxy', where the Super Computer, after millions of years of racking its cogs to the question 'What is the meaning of life?' gives the answer '42'.

I went inside and was introduced to the headmaster and his staff, who took me to a classroom full to the brim with enthusiastic school children, all dressed smartly in their uniforms, with well-scrubbed shiny faces. I was offered sweet tea, which I gladly accepted. Then it was question-and-answer time. How I wished that I'd not had the last pint and done some homework. Then again, I didn't know what questions would be put to me. The children wanted to know all about British history, 'yikes', the Royal Family, 'more yikes', British politics, 'crikey' and strangely enough, Harold Wilson. 'How odd,' I thought, as darlin' 'Arold wasn't Prime Minister at the time. They were especially interested when I told them that I came from Liverpool, where Wilson was once an MP. Now that's a turn up for the books. I wondered why they were so interested in him, unless they knew something that we didn't know? The day went very well, but it was hard work answering the questions which they, without a doubt, had carefully thought out. I wished I'd seen them prior to my visit – I'm sure they must do on TV and radio! I shook hands with everybody and left School 42.

When I hear the expression "he/she doesn't suffer fools gladly," I used to think, 'What a stupid, arrogant statement,

I mean, who does suffer fools gladly?' As Eddie Izzard says: "Hello, fool! Come on in, fool, make yourself at home!" However, I used to suffer when this short-arsed cockney instrument technician used to corner me in the bar and I never had the heart to tell him to eff off. He had done the rounds and spotted the 'fool sufferer'. I was having a drink in the Back Stabbers, minding my own business and trying not to have a nervous breakdown with the Pogues belching out drunken Irish, accompanied by their greatest fan singing out of sync. When, 'Oh shit,' I had caught the person of short stature's eye and quickly looked away, pretending not to have seen him. But too late, I could see his head out of the corner of my eye, moving towards me. I say his 'head' because that was the only part visible from over the bar. "Hi, Scouse." "Oh, hi" - making out that I had only just seen him. He told me that he was married to an ex-prostitute from Thailand. He, of course, was a towering figure there, as they are mostly small and slightly built. The family photos were coming out again for the sixth time. I've got nothing against prostitutes as I think they do a good job and must reduce a lot of rape and child abuse. In my mind, all rapists and especially child molesters, paedophiles, should have been handed over to that other person of short stature, the supreme psychopath, the late Kim Jong-il of North Korea; he could have used them for target practice and testing his weapons. But no doubt his son, Kim Jong-un, will carry on the time-honoured traditions. "This was taken in the garden." "Oh how nice." I hated myself for being so two-faced and using such a boring expletive as nice but they were so dull. At that moment I would have preferred to have been locked in a room, tied to a chair and forced to listen to a long-playing record of the Pogues.

'It is not true that suffering ennobles the character; happiness does that sometimes, but suffering for the most part, makes men petty and vindictive.' (W. Somerset Maugham).

Before I strangled him, I think he recognised my grief, as he tried to persuade me to come down from the bar and remove the noose from around my neck. He continued and I could hear him over my sobs. He was telling me how he was 'in with the local Mafia.' Apparently, he would go into the Crystal armed with 200 Benson & Hedges and a Bottle of Johnnie Walker Black Label and hand them to the ever so grateful mobster. No wonder he was made welcome. I think they made him an honorary gang member, patting his head much in the same way Benny Hill used to pat poor Jackie Wright.

Another character, I nicknamed 'Captain Pugwash'. Pugwash was the most unfunny person that I have ever met, who tries to be funny. One night at the Crystal, he went over the top. To a Russian from this town, a ushanka (Russian fur hat) was one of their prized possessions, as they were not well-off. Ushankas are made from the fur of fox, sable and mink and quite expensive. It is not only a status symbol but it rather keeps your head warm, which is not at all a bad idea when temperatures can drop to -10 to -20 degrees. On the way out, Pugwash snatched one from the head of this huge Russian guy, who could well have been Arnold Schwarzenegger's double in the Terminator movies. Remember the look on the face of the baddy facing Clint Eastward in a spaghetti western? The moment before they draw and the baddy twitches and knows that he stands no chance and is about to die?! Well, that look came over the face of Pugwash – it was the best laugh that he ever produced. Fortunately for him, the Russian was too polite to take offence and kill him but he most certainly got his message over to the Capt'n.

I had some good times in Russia. Occasionally some of the interpreters would invite us to the company's dacha (log cabin in the forest) where there would be a huge log fire, plenty of snow and lots of pepper vodka to warm you up. At Christmas time the ice statues that they made in the parks were out of this world and would not melt overnight like our snowmen in the UK. To play a trick on the unsuspecting townsfolk, as they can be such miserable sods at the best of times (and who could blame them, living in this depressing place?), I would walk out in jeans, T-shirt and trainers. The look on the locals' faces as they approached. They would be wearing heavy coats, ushankas and boots and I would walk past them dressed as if for summer – they probably thought I was crazy. As I got to the other hotel which was only about fifty metres away, I would be freezing but would soon be warm in the Back Stabbers, with all the hot air, and the heating was on too! Funny, I never felt the cold going back – I wonder if it had anything to do with the vodka?

Near Christmas time, I was invited to a local party where there was hardly anything to eat or drink. We had just been to a colleague's birthday party at a local restaurant and on the way home on the coach, I was asked by one of the interpreters if I fancied going to a party. They couldn't ask everybody for obvious reasons: the lack of everything. I arrived at the party to find about twelve Russians with two bottles of vodka and all were sharing cigarettes. I thought, 'Stuff this.' I asked our coach driver to take me back to the apartment gulag to fetch some supplies. We had our own supplier and were also allowed to use the apartment shop as soon as it opened. I grabbed about four bottles of vodka, a bottle of whisky, a case of Guinness, two hundred cigarettes and packets of Hamlet cigars. When I got back I was the hero of the party; in fact, an honorary member for life.

I stayed overnight, as taxis just did not exist in this town, unlike Moscow, where you can be taken just about anywhere you wanted for twenty Marlboro. The next day, we all sat in the kitchen chatting, as best we could through pigeon English and a phrase book. The light was beginning to fade away, when somebody knocked on the door. Everybody just froze and looked nervously at each other. Nobody opened the door. "Why don't you open the door?" I asked. "Shh." And this was Gorbachev's Russia – God knows what it must have been like under that kindly soul Uncle Joe Stalin. BANG! BANG! Everybody jumped. "Jesus, what the fuck's that?" After a couple of seconds, they all started to laugh. It was a neighbour outside in the square beating a rug over a clothes line with a huge pan-shaped object. This eased the tension and we all laughed together. Imagine being at home on a Sunday afternoon after a party, sitting in the dark trying not to attract any attention from the security, being too frightened to answer the door and nearly having a heart attack when someone cleans a rug. And we moan about Blighty; we don't know we're born half the time. Still don't know who was at the door.

We nearly had to break down the instrument analyser's door once, as he hadn't showed up for work for two days and had not phoned in. We got the key to his room from the housekeeper and found him sitting on the floor amongst a pile of empty bottles, spinning a bottle. He was totally gone; I think he'd had a breakdown. Rumour was, he had problems at home. That was him on the next flight out. An analyser that required analysing. That's the second analyser that cracked up – remember the boxer in Iraq? (If not, well, it's in that chapter.)

I made several visits to Moscow and paid a visit to the famous Red Square, which has that fantastic Saint Basil's Cathedral, the one with the iced cream cornets on top. I always regret never going inside. Oh and that other place, the Gremlin, er, Kremlin. You are allowed to go inside the walls, through the Borovitsky gate but you had to watch out for the Zils (politburo cars) coming and going. Inside was amazing - the many churches in there are absolutely superb. Red Square has Lenin's mausoleum and every hour on the dot the guards change over as the clock chimes. They do this most peculiar slow goose-step: their heads goes from side to front and their free arm goes sideways. I don't know how they can go so slowly without falling over. One day the devil was in me. I walked over to one of the main buildings as if lost, to where the politburo does their biz, and was soon escorted away by one of the guards who always look huge (but by the time they take off their hats and their boots, they'd be a mere five foot two). One got the sudden feeling of being in a James Bond movie. Red Square was a lot smaller than I'd imagined it to be, as the only time that you see it is on TV on May Day, with all the tanks, missiles and the bunch of grim bar-stewards in big hats and overcoats looking down as if to say: "Look what we have," in a rather deep sinister voice.

It makes my blood boil when I think of the way people are treated, or I should say mistreated in some of these countries. I mean, what right do they have to frighten, oppress and torture people and govern tyrannically? Despots, zealots, tyrants. Many people in the UK, including me, criticise our nanny-state government, which I must admit are taking things too far, a touch of George Orwell's 1984, but when you have visited some of the places that I have been to, you would think twice before complaining again. I feel as if I could have a pint with Dave or Nick and discuss life, politics and share a joke. Imagine if you had said to Saddam Hussein, "Fancy a pint, mate? By the way, I don't think much of your policies, of starving and murdering people and wasting all the oil revenue on arms and all your palaces. Oh and how's Uday, that murdering, psychopathic son of yours?" You would soon be in an acid bath. Or not being allowed out of your country, having to queue for years just to go to the Costa del Sol – on second thoughts that may not be that bad, but you get my picture. Criticise me or not, but matriarch Maggie, for all her bad points, to some extent put the lion back and got us some respect around the world, as believe me 'strength' is the only word that crackpot despots understand. Saddam Hussein and loopy Gadaffi have now departed. Next, no doubt, will be Bashar (The Basher) al-Assad of Syria. Then we are faced with Iran's mad Mullahs and their president Mahmoud Ahmadinejad and, oh yes, another country springs to mind, good old China and the nuke kids on the block, India and Pakistan. The lists grows and grows and what about here, Russia/CIS, now that drunken Boris has gone to the vodka distillery retirement home? His replacement, Putin, is a blast from the Cold War past, an ex-KGB boss and his replacement, Medvedev, the puppet on Putin's string. So much for perestroika and glasnost! We have got to have a strong foreign policy and must never ever disband or weaken our armed forces. What would be the point if we did? Yes, it costs money and welfare needs money, the NHS needs money, everything needs money, but we are dealing with some psychopathic monsters that kill their own people, children included, and countries that kill young students because they protest and do studenty things. It would give all the world's nutters the green light! Sorry about that, got a bit carried away there and this is written from a 1989 perspective but I ask you, what has changed?

Our project had been going on for what I was told was ten years, as the Russians kept running out of funds and of course the company that I worked for at the time kept pulling out. This showed itself in the state of the badly preserved and neglected equipment. Some of this equipment were hydrogen compressors, compressing hydrogen to 2000 p.s.i. Which, if not commissioned correctly, can go 'pop' in a big way. Hydrogen (H_2) plus oxygen O_2 has a low-ignition flash point and can auto-ignite. On a hydrocracker, there must not be any leaks and an N_2 (nitrogen) purge is required. The O_2 content must be less than 3% and any leaks will cause ignition. I am sure that many people do not realise the dangers of charging, say, a car battery, as it gives off a hydrogen gas which can ignite as it produces H2 and O2. So with such high pressures on the plant, a leak could have a catastrophic effect resulting in a fireball and or an explosion.

We had to commission this equipment after it had been mothballed for some years, minus the mothballs. But before we could commission it, we had to pre-commission it, which meant that one of the most vital things we had to make sure of was carried out meticulously - to flush out all the lubricating oil system. This was done by bypassing the main bearings so that all the debris and bits of iron filings etc that got trapped in the micron filters did not get through to the bearings: all the remaining parts only had minute tolerances: 0.002 of an inch, less than the thickness of a human hair.

It was my job to inspect these micron filters and check they were clear. If something happened, then it would be my head on the chopping block. Go direct to Gulag, do not pass Red Square, and do not collect 200 Roubles. It is a tense moment when you start something for the first time, and a hydrogen compressor, compressing pure hydrogen at 2000 p.s.i. can be a very unforgiving creature – well, you've heard of hydrogen bombs. The Russian engineers would send for me to do my inspection. One day, I went half an hour earlier and caught them doing 'naughty things' like taking out all the debris, bits that had been trapped by the filters, so that they would 'appear clean', expecting me to give them permission to go onto the next stage. I would then piss them off by saying "Nyet, continue another twenty-four hours flushing and I want to be present before you stop the pumps and remove the filter screens." You know, in case the thing explodes. In a hydrogen-bomb fashion.

We would have meetings with the directors of the plant to arrange and agree a time for tests at, say, nine o'clock in the morning and would end up freezing our goolies off waiting all day for the Russian engineers to show up. They would appear at about four o'clock in the afternoon when you were just about to finish for the day. It would be dark as well, so the tests would have to be cancelled. To start up for the first time, you need all the light that you can get, in case of any leaks etc. You have to monitor the equipment very closely for obvious reasons (explained explosively above) and for maybe up to seventy-two hours nonstop. I couldn't understand their logic, but that's Russia.

The control room was old and decrepit and full of cigarette smoke. It is a definite 'no' to smoke inside a control room, because of all the delicate instruments that are taking and recording vital figures etc of the plant. Which brings to mind Chernobyl again. The plant had been hardly used but it was ten years old and was deteriorating before our eyes. It now had to be recommissioned. The Russians would be running around in a panic, trying to get you to check pipelines, valve seats, special high pressure joints and gaskets, most of which were now corroded and scrap, even though they had never been used. I just thought of all these nasty mediums going through them at 2000 lbs p.s.i. 'Yuk.' Of course you did not pass it and sign it off for the next stage but you would never know if they had heeded your warnings and instructions and rectified the all-too-apparent short comings or not, as you could not force them. The plant was producing linear alpha olefins: any leakage would start a fire and give off toxic fumes. Not nice stuff – you had to be careful what you passed!

When on site, we had to carry a dosimeter at all times, as they would start to X-ray welds etc and would not bother to warn any personnel at the plant, nor put up any radiology warning signs (like the skull-and-crossbones type, ones that tell you that X-raying was in progress). If you heard a beep when ionizing radiation was detected and the defined limits of dose rate or cumulative dose are reached, you would have to vacate the area rather smartish! The problem was, you wouldn't know where to go, where the safest escape route was and was it gamma or X-ray? The times that we protested over this issue, they just didn't take any notice. Remember the scenes on television of all those brave Russian volunteers in those silly aprons trying to cover the hole on top of the reactor at Chernobyl? I was glad in the end to leave this godforsaken plant.

The interpreter came with me to Moscow and I bought her a going-away present in the duty-free shop. You know what she asked for? Tampax. Russia! Dosvedanya.

A bit cold in Red Square, Moscow.

QATAR
1991

*(In which I pass a driving test and
a diving test but fail my lesson in Gulf Wars.)*

I was in the garden shed when the telephone rang. It was someone I knew from Bechtel's London office, Bechtel being one of the world's largest construction and engineering companies.

"Are you doing anything at the moment?"

"Apart from gardening, no."

"How does Qatar grab you?"

"Qatar as in Qatar, Arabian Gulf?"

"Yes."

"Probably by the balls at the moment, why?"

It was November 1990 and the Alliance, or rather America, had given Saddam Hussein until 15 January to get out of Kuwait, which his forces were now occupying.

'Are you a man or a mouse?' they say. Well, I have always maintained that whether a black cat crossing your path is lucky or unlucky depends on whether you are a man or a mouse (squeak). I like to think that I am a man but when it comes to being bombed and gassed, I tend to get a little mousey.

"We are behind with our pre-commissioning and commissioning schedule."

"I bet you are."

"It will be worth your while."

I was in-between contracts, not a good bargaining position to be in. If you turned down a contract, you could not always be sure when the next would come along. As the fat guy with white beard and red suit was heading my way fast, I agreed.

Arriving at Heathrow, I found myself feeling a wee bit apprehensive when going through the gate. 'Am I crazy or what?' kind of went through my mind. Having already been involved on two occasions with the Gulf Wars, was I pushing my luck a bit? Third time, unlucky? It'll be OK, I told myself, it's not due to kick off until the middle of January anyway. Enough time to get in, suss the place out, get a feel of the situation and if things started to go against the Alliances, sorry, America's wishes, then I could exit stage left and get the hell out of there.

I was taken straight to the office where I went through the usual routine of meeting people and filling in visa application forms, certs checks, having a medical, giving blood samples and the obligatory handing over of the passport, which was the most unpleasant of all. The accommodation was good but it was about forty-five minutes away from Qatar itself. So I would need a car. Not easy to obtain. One of the first things I noticed was the amount of military aircraft on the ground at the airport. Never mind, at least it wasn't a dry state and you could have a drink. The setback was: you required a visa to get a liquor licence and it took a month to get one. It was going to be a long wait...

The field office was the open-plan type and my desk was facing Doug, or 'Bluto' (Popeye's arch-enemy in the cartoons). Bluto came from the Isle of Man and was a rather large chap, a fat, coarse bastard to be exact, hence the nickname, or just plain 'Blute'. You would be sat at your desk, the door would burst open and in he would come, crash his site helmet on his desk and wipe the sweat off his brow with the back of his hand. As his forefinger was the final point of contact, he would then flick the sweat up the office, then slouch in his chair, not too unlike a beached whale. (Sorry, whale.) Then he'd give out this huge belch, and look at you as if you hadn't noticed he was there. If you didn't acknowledge him, I'm sure he'd have jumped onto his desk, beating his chest like a fucking Gorilla. This would be followed by a series of huffing and puffing, then to top it all, or I should say 'bottom it all', he'd lean to one side at a 45-degree angle and coerce out a whopping great fart!

"Oh hi, Doug, hadn't noticed you come in." Instead of what I really wanted to say: "Why don't you eff off, you big fat ugly ignorant uncouth c*nt." Or words to that effect. When you thought things couldn't get any worse, out came the Marlboros, followed by a volley of abuse:
"Fucking lazy bastards."
"Who's getting your seal of approval today, Doug?"
"Fucking planners."
Blute was an ex-submariner, so he was used to sinking to great depths. He was a proverbial troglodyte. This was my signal to postpone my paperwork and to go on site.

The day for me to take my driving test came as a British driving licence was not accepted. I had heard about this test from my colleagues and corpulent Blute, who I assume was hoping I'd fail. I was dreading it. The police compound was full of the usual Indians, Filipinos and young, rich Qataris waiting to be unleashed in their 4x4s, Porches or Ferraris. I filled in the obligatory forms and paid my money. As you never usually passed at the first attempt, it was a nice little earner for somebody, AKA the chief of police no doubt. My turn came and I was taken to the test area. It was a small circuit, surrounded by high wire nets. As I appeared, masses of turbaned chaps and Filipinos gathered around and were climbing the wire nets to get a better view, which was rather off-putting to say the least.

You had to drive around the circuit in reverse and then reverse into what was supposed to represent a garage. I felt a thousand eyes staring at me, willing me to hit the fencing, which I did. 'Shit.' That meant a retake in a week. I had to get my licence as otherwise I'd be stuck in the camp. The week seemed to be going in reverse too – it dragged like hell. On my return, I managed the reverse section and was wondering what was next. I had to drive on the road, with two Qatari policemen in the car, one in the front and the other in the back. Despite this, I passed. You then had to be fingerprinted at the police station, another little earner for 'you know who'. 'No problem,' you think. Ever tried to negotiate your way round a very crowded room, full of people covered in black ink? There were no washing facilities, paper towels, nothing, and I was wearing a white shirt. I couldn't wear it again as I didn't want to be mistaken for a Dalmatian. In walked this horrid fat Qatari woman, dragging behind her a small Indian guy, who would probably be the servant/slave? She was smacking him on the head, kicking him and shouting at him. God, what a life and what a cow! These poor sods work for peanuts, seven days a week, and probably never see their families for one maybe two years at a time. She was treating him worse than a dog.

One of my office colleagues invited me to a beach barbecue at the weekend. Actually it was a B.S.A.C. (British Sub Aqua Club) diving weekend. I'd never given diving a thought before and after watching all the divers donning their equipment, I started to feel a bit (a lot) envious of them. So I borrowed a mask, snorkel and fins and followed them out. It was a shore dive, which meant that they would walk out to waist depth, put on their fins and gradually descend. Looking down at them in another world got to me.

"How do I join?" I asked when I got back to the beach. "Well, for starters, put on that weight belt and let me see you swim a hundred metres, without stopping." This I did. "Next, when we get back to the camp, I want you to dive in the deep end of the swimming pool and pick up lead weights, then tread water with your hands behind your back for two minutes." After the barbecue and drinking one or two beers, I found it somewhat taxing but I did it. "Great, I see no problem. Now for the hard part. As you are two weeks behind the other new members, you're going to have to do a little cramming to catch up."

The first lecture was a little gruesome. It was Boyle's law: how your lungs could burst like a balloon if you surfaced too quickly. Good old Boyle's law. You got all the nasty stuff first: the bends, the narks (nitrogen narcosis) where you feel happy when you're really in deep shit and are inclined do all sorts of crazy things such as kill oneself. I didn't much like the sound of them. Then there were stab-jackets (stabiliser jackets) and knives, handy for cutting yourself free from old abandoned fishing nets. This rather spoilt the myth about fighting sharks but you'd be surprised at the amount of these nets you can encounter on a dive, fatal if you did get trapped and had no way to free yourself. And of course, fins (rookie divers would call them flippers and the more experienced divers would correct you with a smirk). We then had lessons in the swimming pool, where we were shown how to use the diving equipment, put it together and then how to put it on. Then do all that underwater. Oh and how to get in and get out of the water with it all on.

"You want me to do what?" This next bit involved swimming underwater blindfolded, without your mask, trying not to breathe through your nose, only through the DV (demand valve) in your mouth, with of course the air tank on your back. Then you and your buddy (dive partner) would swim alongside each other, sharing one mouthpiece. You'd have to remove your mask underwater and put it back on again and clear the water from it by looking up to the surface, applying pressure to the top of the mask and then blowing like hell through your nose. The water should then come out of the mask, as the theory goes. I hated this. I've always had problems breathing through my nose. It may sound simple but try doing that on the seabed with murky, stinging, cold, salty water whilst you're unable to see.

This was my Achilles heel. Clearing your mask is a very important procedure which you had to master but for the life of me, I couldn't clear it 100%. Now, all the instructors just had to pick this up, didn't they? I didn't mind a little water in my mask as I could swish it around to clear any misting. But you could be on a deep dive with a crowd, groping around a wreck, in poor viz (visibility) that could be down to a metre and someone could inadvertently knock your mask off or dislodge it with their fins and you'd be in serious trouble and, to be honest, more likely than not to survive! So what did they do? Yes, they made me repeat it over and over again. I'd be on a training dive and an instructor would come from behind and snatch it off my face, letting it fall to the seabed. Through sheer panic and with the sea stinging your eyes, you'd have to retrieve it, put it on and clear it. This would be repeated until I'd mastered it. I didn't quite see the funny side as I think I was traumatised at the time but I appreciated it eventually as it came in handy on later dives.

There was this Welsh guy in our diving club in his late twenties, who absolutely adored himself. He couldn't pass a mirror without kissing it. He'd be in the gym every day and even had a hairdryer with him. I won't go on as I think you know the type and yes, he turned out to be a right shit. We were going to our weekly pre-dive meeting and he was late. I went to gee him up and there he was, blowdrying his hair like a big girl's blouse, in front of two Indian cleaners. God knows what they were thinking. God's gift would sit and snigger at people's misfortunes, which got up my nose. He would later come unstuck...

I passed my novice exams and it was into the sea for my first open-water dive.

"Where's the dive site?"

"Snake Alley."

"Err, Snake Alley as in sea snakes?"

I remember David Attenborough saying: "A bite from one of these snakes will kill you as there is no known antidote." Great.

"Why Snake Alley for our first open-water dive?"

"So we can get rid of all you novice divers."

I was already shitting myself with the prospect of the dive but the idea of diving amongst highly poisonous snakes didn't fill me with joy. I'd been told that they only have small mouths that can only open wide enough to bite you in-between your fingers.

"As they surface to breathe, just pull their tails."

"Oh sure," I said, putting on extra gloves.

The dive started off disastrously and got bloody worse. I looked and behaved like a hapless twit. I didn't have the correct weights in my belt, and I was too buoyant. You'd not be inaccurate to think I was snorkelling and not scuba-diving, as I spent most of the time near the surface and completely forgot about sea snakes. The afternoon dive was, er, different. This time I had too many weights and kept scraping the seabed. It took me a good few dives to get my buoyancy correct. You're supposed to hover like a fish, not a sea slug sticking to the bottom. When you get it sorted, it's a great feeling, like you're flying. I immediately saw why divers carry such large knives as part of their equipment. Abandoned fishing nets cover the seabed like giant spiders' webs, waiting to trap you, and if you did get caught up in one and didn't have your knife to cut your way out, it'd be curtains, or in this case, nets.

The diving instructors had gone and we novices were still in diving mode. "Come on, let's do one more dive," some wise guy exclaimed. I was still feeling a little euphoric after completing my first open-water dive and, as we had about an hour's air left, we got back in again, breaking all the rules: novices should not dive without a qualified and experienced diver. We didn't have sufficient air left in case of emergency, we didn't have a buddy-partner each and we were behaving in a disorderly manner as we were high from the last dive. It was an accident waiting to happen. It was also getting dark and we were unfamiliar with the terrain; we didn't have an oxygen tank in case someone surfaced too quickly and we didn't have a clue where the nearest decompression chamber was, if there were even any in the country! Also, we had no telephone, no dive boat, and last but not least, no diving plan. The divers' favoured saying is: "Plan the dive, dive the plan." Did I mention this was an accident waiting to happen?

Again, I had buoyancy problems and whilst arsing about with my stab-jacket, the others, more advanced than I was, were off fast. I soon lost sight of them, not good for an experienced diver but a bloody nightmare for a novice, believe me. I waited for the required two minutes for somebody to return but nobody did. It sure was a long time. All sorts of things were going through my mind: sharks, sea snakes, giant spider fishing nets… As you can imagine, I was in the shit. As I didn't have a compass, the only thing to do was to surface. Just as I was about to send air into my stabiliser-jacket to attain lift, the sun's last rays of the day shone down on to what could only be described as a boxing ring. I thought I was dreaming. There, on the seabed, a fishing net had somehow shaped itself into a square and the floor of the square was sand. The effect of this, with the sun rays beaming on it, was really surreal! I surfaced and found that I was only about two hundred metres from the beach. The current wasn't too strong but I didn't give it any time to build up. I rolled onto my back and swam to the shore.

The final practical test for my novice diver's certificate was a deep dive and I wasn't feeling too good after having one too many drinks the night before. I asked the dive leader if I could put it off until a later date but he told me it might be another year before I could take it again as the examiner may not be available. So, waiting to get the novice certificate out of the way so that I could get my sports certificate, which enables you to dive without an instructor, I agreed to take it. "I'll put you on the last wave (the last dive)," our dive leader told me, "give you time to get yourself right." We set off on the rib and I wasn't at all sure if it was a good idea being last down, as you not only had the apprehension of waiting, but also the fumes of the boat's two-stroke engine to contend with. The two divers before me surfaced with their instructor and my heart started to race: this was a thirty-metre dive which is deep for a novice diver. This was it, no way out now.

I did a backwards roll into the water and the instructor gave me the diving OK signal, I reciprocated the signal and we descended down the shot line that was attached to the boat. I seemed to be going down forever; you could feel the difference in pressure every metre of your descent, as your face mask started to squash your face. You then had to pinch and blow your nose to relieve the pressure to avoid mask squeeze, otherwise you'd end up the next day like a panda. 26, 27, 28, 29 and bump, 30 metres... landed. It was a most weird feeling as you approached the seabed at such a depth, like landing on the moon. A total alien world. To my surprise, I felt really confident and at ease. I knelt down on the seabed and checked my equipment, making a few weight-belt adjustments. The viz was only three metres. We tied a line of string to the anchor and swam off. There was a brief moment of panic when it dawned on me that I was almost 100ft below the surface but I said to myself, 'Bollocks,' and carried on.

We exercised a controlled ascent to the surface, got on the boat and gave the two Qatari divers high fives, then headed back to the beach for a well-earned beer. As we approached the beach, I felt like James Bond or the S.A.S. as there were so many people watching us come in but this feeling soon evaporated. As we came to our part of the beach where our diving crowd was situated, I spotted a police car. 'Allo, allo, what's going on 'ere, then?' We had some bikini-clad female divers with us and this blew the young Arabs' minds. When you think that Arab females bathe with all their clothes on, the sight of bare thighs, midriffs and breasts almost falling out was just too much for them. We had set up our beach camp about twelve feet from the water's edge and the local Qatari males would drive their 4x4s between us and the water's edge to leer at the women. This had pissed off our dive-club members as some of the divers had their children with them. We arrived just as the police car moved off, with Welsh Boyo in the back of it and this time he wasn't smirking. His dad, who was a really nice guy, said the Qataris had come too close and his boyo had struck out at one of them. I had to stifle a grin but he'd kind of stolen our thunder for what should have been a good celebration. Our dive leader said not to worry, he'd be out in an hour or so. 'Only an hour?' Damn.

The morning meeting was depressing as we learned that none of the vendor engineers would come out to commission any of their equipment. This left us in more trouble. We needed them as, obviously, they were the experts on their equipment but they were quite more concerned with the impending war. "Sod it, let's do it ourselves," I said. Our French commissioning manager liked the idea. "We have enough experience between us to carry it out." We decided to split the areas up into sections. I got the emergency diesel generator, the air compressors and the diesel fire and process pumps. Bluto was becoming a pain as he just wouldn't sign for anything after witnessing the tests. He was taking vibration readings from hydrogen fin-fans which are housed in a special edifice, basically made of steel girders. You could never get a reliable vibration reading from that, as the whole structure would vibrate so much that the needle would go off the monitor, especially as the vibration-monitoring equipment he was using was for more intricate machinery. "They can fuck off. I'm not signing for that shit." The main contractor was an Italian company who were not amused at his antics. If war was not imminent, I'm sure they'd have put out a contract on him. Funny, they kept looking at his shoes… Sizing him up for concrete boots?

Apart from Bluto's farts and belches, I was quite enjoying myself. The work was interesting and the day would go quickly but the weekend couldn't come quick enough for the diving. I had to find something to do during the week after work to occupy myself so I paid a visit to the culture section of the British embassy and they gave me the address of the local drama group. In my teens, I had a part-time job at the Royal Court Theatre in Liverpool, which got me interested in the theatre. The theatre group in Qatar were desperate for a sound person, so I thought I'd give it a go. We had just six weeks to pull it all together and they hadn't yet completed the auditioning. The group were putting on 'The Wizard of Oz' for their Christmas panto. We all met up at director's house and watched the video of the original movie. I made notes of what sound effects were required, such as animals, farmyard, storms, wind and thunder - I thought of asking dear old Bluto for assistance! I ended up phoning the local radio station and spoke to the DJ, who kindly loaned me some BBC radio sound-effect records which were perfect. I didn't need Bluto after all: I just had to transfer it to our sound equipment. I say 'just', it was a mini nightmare as the equipment hadn't been used in over a year. It was all over the place and tangled up and, to make it worse, I'd not had any experience on their equipment before. I had a week to sort it out before rehearsals started.

In work, we were coping without the vendors, which in fact made things more interesting. I was in the theatre three nights per week, one night was a pre-dive meeting and the weekend was the dive and barbecue. Life was busy! The panto went down a treat, we even got a good write-up in the local news. Everything was rosy until I received the dreaded telephone call from home. My eldest daughter had been in a car crash and was in hospital. My son had already suffered serious brain damage at birth, due to a doctor's pathetic negligence. (Not long after, this doctor was promoted.) I thought: 'No, not two.' I was on the next flight out that night and went straight to the hospital. She was lying flat on her back. My daughter and her classmates had been driving along in their school lunch break, had stopped at traffic lights and a bus had hit them from behind. Her head was placed between sandbags for a week as her vertebrae had gone out of alignment and they feared a spinal injury but mercifully she was OK. She made a swift recovery and was soon back at school. When I think what could have happened, it sends shivers down my spine!

I flew back to the Gulf on 14 January and was on the last flight into Qatar before they closed the airport as 15 January was the deadline the Alliance/Americans had given Saddam to get out of Kuwait. I had a quick lunch in the Ramada hotel. The surroundings seemed very eerie and the waiters were silent and acted unnaturally. Then it hit home... What had I done?! A feeling of fear gripped me. What in God's name had made me come back to what could be a war zone? In striking distance from Iraqi Scud missiles, armed with biological and chemical warheads? My appetite soon evaporated. I took a taxi to our camp, dumped my bags, got into my car and drove to our office, where I received a bollocking from my boss for coming back. "Wouldn't have missed this for the world," sounded a real dumb thing to say at the time. My mate, Geoff, came over. "Ken, we've all been issued with gas masks from the British embassy, you'd better get your arse straight there and get one, in case that mad cunt starts firing his Scuds at us." I shot to my room and grabbed my diving air bottle and mask. I had about thirty minutes of air still left. If we did get attacked, I was, at least, in with a chance. I drove into the embassy compound and was given a gas mask by one of the staff, who told me that the foreign secretary, Douglas Hurd, had paid a visit the day before and had given them a talk. He told us that "war was imminent as Saddam was not going to pull out of Kuwait."

15 January, 11.55am. We all gathered round the radio to listen to the world news. The minutes ticked by and 'Bong' midday. The deadline had passed. We all looked at each other, not knowing whether to laugh or cry. The apathy was soon demolished by a huge fart. Well, that's what Bluto thought about it. It was the most welcome fart that I have ever heard. We all creased up with the sudden release of tension, as was Bluto's face as he tried to release another Saddam blaster from his inexhaustible reservoir of stockpiled gasses. Should we tell Douglas Hurd? Saddam has his Scuds, we have our skunk - Bluto, our secret weapon.

Lunchtime came and we had a meeting with our management, who told us that we were all safe and that we'd be on a special war payment. Where have I heard all this crap before? The next day, I overslept. Usually the tea boy would wake you and bring you a cuppa but he was a no-show. I got out of bed and put the BBC World Service on. One of the newscasters was saying: "I am standing here in the centre of Baghdad, watching a cruise missile heading towards its target." Here we go again, I thought. Had the third time come? 1: Basra, Iraq, the start of the Iran/Iraq war. 2: Northern Iraq, during the Iran/Iraq war, and now 3: Iraq and Kuwait. I quickly got dressed and called next door to my neighbour. We both sat, transfixed at the radio. Wave after wave of cruise missiles rained on Baghdad. I swore that I heard the newscaster say he saw a cruise missile stop at traffic lights, signal and turn left. Well, not quite, but one day they'll have this capability - well, not to stop at traffic lights but stop, hover, have a think and change direction. Tremendous, just what we all need!

That night, I drove into Doha. On my way I stopped at a small supermarket and noticed straight away how different the locals were. They seemed hostile towards me and I didn't hang about. As I approached my car, the sky was suddenly lit up by F16s' landing lights. They had been on a sortie. The sound was awesome in the old sense of the word, the formidable might of the Alliance, er, American air power. This was too close for comfort and brought back frightening memories from the past. I headed back to the camp. Some of the odours wafting from the refinery didn't help the nerves as the thought crossed my mind of Saddam's Scuds delivering his chemical and/or biological weaponry. Some of the advice given out by the British embassy was: 1) Observe outside events: if people or animals start to act strangely or fall over. (What's new, as most of the people in our office acted like that anyway?!)

2) Notice strange odours. (As we had to put up with Bluto and his strange odours all the time, how could we tell? Most of the guys in the office took to wearing their gas masks because of him anyway. And I must admit, they looked better with them on.)

3) Use your office as a 'safe haven'. Seal all the doors and cracks. Turn off the air conditioning. (Obviously, whoever wrote this does not have to share an office with Bluto.)

Then we'd get the ludicrous TV news, where you could follow the missile to its target. Some people started cheering. I went to my room and poured myself a large Scotch and lay down on my bed feeling dreadful. I looked around my room. My stabiliser jacket was draped around a life-size cardboard cut-out of an air hostess that I'd obtained from the local travel agent. My gas mask was on top of the wardrobe looking down at me. It was such an abominable thing. I sipped my Scotch, looked at the mask, it looked at me. I gulped my Scotch and said, "Fuck it, I don't need this." I decided to leave the next day. The company had given us the choice of: 'Er, staying or going,' which was jolly decent of them. As I'd survived the last two Gulf Wars, I felt I shouldn't push my luck any further. Besides, it wasn't fair on my family.

The next day we headed off at daybreak. Our convoy consisted of three vehicles and part of the journey was through the Saudi desert. All was going well when one of the vehicle's engines started to overheat. We pulled over and I had a look under the bonnet. Out of a series of V-belts, one had broken and it was the one driving the engine-cooling water pump. Of course the driver didn't have a spare one: he had bought this new 4x4 in Qatar and was taking it home to the UK. Fortunately all the belts were the same size. I removed the belt that was driving the air-conditioning unit and put it on the cooling water-pump pulley. They could survive without the AC but the engine would not survive without cooling. I couldn't believe the heckling I was getting from my ex-next-door neighbour. He was severally criticising me for helping the poor guy with his brand-new 4x4. It took me about half an hour and he wanted me to abandon it. It probably cost the guy six months to a year's salary. I cannot, for the life of me, understand some people. A real selfish bastard. (If the guy whose vehicle it was is reading this, you know who you are and you still owe me a drink.)

We arrived in Dubai late afternoon and checked into the Chicago Beach Hotel. Later, we all met up for a meal and a drink. An older Welsh guy (not boyo), with whom I'd got on very well during my time in Qatar, tried to pick a fight with me. He'd misheard something I'd said and wouldn't listen to anyone when it was explained to him! It shocked me as he was, I thought, a mate who then turned out to be a complete arsehole. He apologised to me the next day but I found that I couldn't take to him after his outburst. You feel so let down. I never set eyes on him or heard from him again. Oh well, circumstances take their toll on fragile friendships made and lost in this line of work – it's the price you pay. The flight was an extremely long one, as we had to fly south over friendly countries. We skirted Oman, then along and up the Red Sea, past Egypt. It wasn't the best way to end a contract, but at least I ended up a qualified diver.

CHINA

(In which I very much change my mind from when I was a boy.)

1. Shenzhen

1994

When I was a boy, I was afraid of Chinese people. My older sisters drilled fallacious thoughts into my young impressionable skull: 'They're weird, evil, sinister, like Fu Manchu with long beards and opium!' It didn't help that a Chinese fish-and-chip shop opened at the end of our road and, being the youngest of our brood, I was tasked to go and fetch. It was a huge step to take as it seemed like going into the devil's den, and to make it worse, I couldn't see over the counter and when it was my turn a strange face appeared looking down at me, speaking in pigeon English. I wanted to run out of the shop. My devilish sisters then tortured me with having to eat this 'evil food from the devil's kitchen'. Even though I was starving, I hated the thought of it. Now here I am telling you about my trips there. How things change when you grow up and learn.

I flew out to Hong Kong and it seemed to have changed quite a bit since I was last here, certainly since we Brits vacated the place several years ago. Now, it seemed more militarised and communist. But I was here, the great mystic East. First thing was to get into a hotel. Considering it was now communist

China, I went through customs without much of a fuss, just stern faces. The next day, I headed in a taxi to the harbour to board a boat to Shekou, Shenzen in Guangdong province. I was to spend the next month or so commissioning an oil rig for Philips 66 Petroleum Company (I believe that they got this name from the famous Route 66, although how true that is I am not too sure). I remember that first footstep in China, stepping off the boat, which seemed quite strange and exciting. I was looking forward to this project, who wouldn't?

Arriving at the office of Philips 66, I was introduced to all their staff, all Americans with big hats jeans, boots and belt buckles - typical Texan oilmen. After several meetings that day, we headed back in a minibus to our hotel. I was with the company's management and, to my astonishment, as we overtook another oilmen's minibus occupied by our office staff, several of the managers started banging on the windows and hollering at them. To my utter amazement, they dropped their pants and mooned at them with their arses out of the windows. 'Well,' I thought, 'this is definitely not British Petroleum's way of conducting themselves!' But I was here for the ride, oh and the cash as well. After working on BP's projects with all their stuffiness, I was ill-prepared for this. But sod it: as ever, just go with the flow.

The first week was to be spent in the Shenzhen office to get up to speed on the project, which was their rig located in the South China Sea. After an unfriendly episode at the local immigration office, where I had to register and where the officials were very official indeed, my visa application was granted. After dinner that evening, I was invited to go for a drink at the Snake Pit, the dreaded local bar. I couldn't help but hear some tittering within our group and, being the new boy, I was waiting for the ubiquitous 'initiation ceremony' and indeed it came. The barman brought this bottle over and poured out a glass. Of course the bottle just happened to have a dead snake inside, what else?! So I was obliged to knock it back in one go. It was so strong, it evaporated before it reached my lips - and before I was accused of cheating, the law invaded the place, manhandling the local drunks and prostitutes and saying something to us Westerners that was definitely not: "Welcome to our shores!" but more like: "Sod off to your hotel," but not quite as polite. We promptly adhered to their request.

The next day, it was to the heliport to catch the chopper out to the rig. The accommodation was on a 'flotel' - a ship mored next to the rig, which doubled as a workshop. The room was OK but I had to share with this American who pressed his jeans so hard they would have come in handy cutting bread into slices. I quite like Americans as they can be quite odd but then I'm British and definitely a bit odd myself, so you plough on.

The main job was to commission two large marine diesel generators, which was easy enough but the hard part was sharing the toilets and showers as the locals tended to use them – to put it politely – like a hole that peasant farmers would use in a field. It was awful and I made a mental note to always try to get there first. The first weekend, we had to chopper ashore as a hurricane was heading our way. An unexpected treat, a weekend in Honkers and all paid for. We couldn't do much in a hurricane as they can be a little naughty and this one was. After the storm, it was back to the rig. I picked up a week-old newspaper and couldn't believe what I read. The great Ayrton Senna had been killed in a Grand Prix race! Being a huge motor-racing fan, I didn't feel like eating after that news and had an early night.

The days started with the ubiquitous meetings followed by the ubiquitous stormy weather as I was beginning to realise was custom in the South China Sea at that time of year. It didn't take too long to commission the two marine diesel generators and so I was heading home again, back to Blighty. I was in the frequent-flyers lounge when a dozen or so contractors from the rig barged in, all pissed and making a real show of themselves, swearing and getting more drinks as it was all free. I hid behind a newspaper dreading that they would recognise me; fortunately they didn't. The trip was pretty uneventful apart from that! I hadn't experienced the real China but this would come at a later date...

2. Beijing
2007

This project was with an American company called Apache. Strange name for an oil company but that's America for you! It would have suited an outdoor-adventure store much better but as I received my pay cheques from them, what's a name anyway?

So, it was on to yet another white bird, British Airways. The first task was to find a certain hotel in the centre of town, where I was to meet my new colleagues. Some task as Beijing is not the smallest of cities around the globe. Fortunately, the taxi driver spoke some English and I had a name and the address of the hotel - a name I cannot recall but it was 'something Palace' as most are. I soon met up with one of the crew - a Taff, a Welsh electrical engineer from Llan-somewhere-or-other in the valleys. I was soon settled down in the hotel lounge bar and brought up to speed on the project, which was offshore on an oil rig in the South China Sea. We were to spend the first trip at the head office in Beijing, which was good, to enable me to familiarise myself with the project.

The office was huge and open-plan, which I hate. Taff and I shared a corner so we could have a good natter during the day. One of the first things for us to do was find good accommodation as we didn't want to be staying in a hotel with its piped music and staff being over-attentive, especially to a grumpy sod like me after a trying day! We had to involve our creepy agent, a gobby Canadian guy, who was more worried about his profit margins. We soon found an ideal place, a 3-4 star apartment/hotel, with staff who did cleaning and laundry but didn't breathe down your neck every minute. It was spot on and it had a bar (we didn't mind the attentiveness of the bar staff). It had a nice 'home feel' and was somewhere for Taff and me to have a private conversation. So, after work it was straight out of the taxi, into the bar to start pulling the enemy to bits and make plans of how to execute the job!

The day would start downstairs for breakfast, then reception would book us a taxi. The Beijing taxis were not too bad and a decent price but they also had these weird little motorbike taxis, with what looked like a dog kennel on the back, which we did not use. Well, I know I didn't use them but Taff, not so sure about, as he liked to visit a bar at the weekend (whose manager was a Taff also – two Taffs!) which showed non-stop rugby on the TV. So I'm not too sure how he got back after an afternoon on the local brew and Welsh rugby songs. I very much doubt that he would have fitted in a dog-kennel taxi. I once made the mistake of giving the taxi driver old Chinese money instead of renminbi, which did not go down too well with him as it was no longer in circulation. We later found out that it was less than worthless (oops, point taken!) but it was quite strange to see these strange little vehicles with people looking out of the grill. I could have sworn they were barking... well, barking mad riding in them as the traffic was an absolute nightmare and the pollution was awful, as you couldn't see much and could taste the foul air.

Outside our office were several female security guards, very military and no smiles, even after our corny jokes and attempts at leg pulling. I don't think they liked Scouse or Welsh humour. Last in made the first coffee. Even Taffs can make a decent brew, as the coffee and milk was Nescafé and in little packets and Boyo's first task was successful. Soon, the local staff would arrive. I had a right-hand man, sorry woman, an engineer named Angela who fortunately spoke good English and was very clever which helped no end. We would attend the morning meeting, then return to our computers to find the many queries waiting for us to resolve. Things were breaking down and/or malfunctioning on the rig and when an oil platform/rig shuts down, lots of money is lost, so no pressure!! It could be quite demanding, as you had to be careful with the translation so thank God for Angela. My next task was to chair meetings with our contractors, basically having to pull them into line, find out what the hold-ups were and where the money was going - never a pleasant subject. 'Raise voice, bang fist on table' was not the way as, like most people, the Chinese don't like to lose face. So other devious (practical) methods were employed and must have worked because A, they got back on schedule again and B, I didn't disappear or get murdered.

It was February and Chinese New Year. I had dinner in the hotel, then went to my room with a bottle of red wine, opened the curtains of the very large front window, poured a glass and watched the firework display, which was a cracker, so to speak! After all, the Chinese did invent them. It went on for most of the night; even the poorest of people would spend lots of cash buying the fireworks and celebrating New Year. There were some bleary-eyed locals in the office the next day and of course many missing too. I must admit that first coffee went down a treat, good old Taff. I was called into my new boss's office (who turned out to be an inexperienced arse) and told to fly to Shanghai with Angela and her manager for a meeting and to inspect some equipment with our contractors. We arrived in one piece and checked into the hotel. I had made prior arrangements with my old mate Greavsie whom I'd worked with in Africa twice and in Singapore and had been on the same project in Venezuela at a different time. We had arranged to meet in a bar in downtown Shanghai where he was based - no problem for him but a bit different for yours truly. Somehow, I made it there but the bugger was late - probably been eating as he loves his food does. Guy. We ended up in some old joint. I didn't drink too much as I envisaged getting back to my hotel and Shanghai is not the friendliest place to be at night-time. I managed to find my way back, to find a note with my key, from Angela: 'We must fly early next day for another meeting and to inspect some more equipment for our oil field, see you at breakfast 5am.' Shit! I'd just had coffee. Fortunately, we got a call from the contractor cancelling the meeting until the next day, so I was down for a sightseeing tour of Shanghai. Was I lucky?

With our very own tour guides, we started off at the bund, the sea wall, then visited an old shopping area where we had a gorgeous lunch of dim sum. A free touristy day was had. At check-in the next day, I saw a huge advert for cheap life insurance which didn't give me the best confidence on the unknown, small, Chinese, internal airline taking us to a town in Hunan province. After purchasing my death policy, we boarded the flight. Obviously I never made a claim as I am writing this! But we got there and a driver picked us up and took us to our hotel which was decorated with waterfalls and flowers. I just had time to put my bags down, when we were whisked off to the workshop of one of our contractors to carry out inspections. This was soon interrupted for lunch, which just happened to be a banquet, what else? We all sat down around a huge table. Most of the dishes I couldn't recognise, so I asked the girls and was immediately sorry that I bothered as there were some strange items on the menu. I won't mention them as you may be eating. I tended to go for what seemed to be vegetables and even the bloody vegetables

I could swear moved on the plate but I didn't want to let Blighty down but show the Chinese that us Brits were tough. I got the feeling, however, that when I swallowed whatever it was, it swam back out again.

With the lunch banquet over, we resumed the meeting, which was not the usual type of meeting that I'm used to, as obviously it was in Mandarin. Luckily my English-speaking engineer colleagues were with me but believe me, you don't half feel a fool being at a meeting trying to find out what the hell was going on and having to write a report on it too. Taff would have been at home, though, as – no offence – but Welsh is a bit Chinese, isn't it? Well, to the English it is. My expanse of Mandarin is 'hello, please, thank you and goodnight'. The trip, I believe, was a success and we managed to do what we set out to do, so I was glad to be in the lounge bar at our digs with Taff plus a large red wine in my hand, even if some of the conversation was in Welsh. And of course Taff had been run off his feet, hadn't he? And pigs can fly and Taffs can lie!

At the next morning meeting, we were sat waiting for all to arrive when in walked one of the American managers. An English technician sitting next to me said under his breath, "Fuck me, it's Widow Twanky." The American looked like a pantomime dame as he'd obviously had a cosmetic operation to remove the bags under his eyes. The procedure would be less expensive in China and looked it too! The stitches looked like mascara had been applied by a blind person with a half-inch paintbrush and matt-black paint. I couldn't hold it in and just burst out laughing, my coffee going on my open notebook. It was pure slapstick but all the Yanks kept a straight face! Can't beat the British sense of humour, can you? Needless to say he didn't have much to say to me after that.

The weekend was soon upon us and we had planned a trip to the Great Wall. It was freezing cold on the day, as we got off the bus at a place called Badaling and were immediately surrounded by souvenir-hawkers trying to sell their trashy bits and bobs. One dared not look at them because they would think they had you in their clutches. The wall, the part that we walked along, seemed to be new and like it had been rebuilt, which I am sure it was, as it's the most visited part. We managed to get to the chairlift to take us out of the way of the hawkers and I managed to get a photo taken with two forbidding-looking guards in their uniforms armed with swords and axes. Of course I paid them for the privilege, as I didn't want my head displayed on top of a spike.

The ride down was a wee bit scary as you sat on what I'd describe as a tea tray with two levers that you steered with (or at least tried to!) down a steep chute. One of our bunch managed to up-end, probably got his left mixed up with his right. He survived but it could have been worse for me as I'd given him some work to do, to get my own back on his Widow Twanky remark. The bus stopped at a cloisonné workshop – clocks and ornaments made out of metal and all hand-painted, beautiful stuff but too heavy to take home (well, this trip anyway). My wife came out for a visit and did a lot of sightseeing in Beijing with me at work. I took some time off for us to get around, starting with the silk and pearl markets, then to see the Terracotta Warriors which was an amazing place. There were hundreds of them and we actually met two of the farmers that discovered them whilst they were digging in a field to grow crops. Their days were now spent in an office getting their photos taken with you and of course receiving cash from yours truly. What a find, eh? From digging to collecting!

Next was a cruise down the Yangtze. We'd booked a first-class room as it wasn't too expensive but got a below-decks room and the dining area wasn't quite up to expectations either but it was only for one night so we put up with it.

We went up to the bridge and had a chat with the captain and I gave him a pair of binoculars for a go at the wheel of his ship, so I can now say that I have driven a ship down the River Yangtze in China! Fortunately, it didn't run aground or into another craft but a wonderful experience. We then got conned coming out of the Forbidden City by two local students. They obviously wanted to get some English-language experience, we thought! They took us to a place where they just served tea – I don't mean like a Joe's cafe but a ceremony, trying different types of tea from all over China. Lovely. Until we were presented with the bill, which was ridiculous. I should have paid more attention to the Visa and MasterCard signs on the walls. Plus, we had given the two students a good tip each as well. I was even more of an idiot as I'd read about this scam in the Sunday Times but it ended well as I refused to pay the whole bill, halved the students' tip and now drink nothing but coffee.

Our next experience was a trip in an hot-air balloon which we'd booked at the hotel. As we got there, they were inflating it. We climbed into the basket, along with a Russian family, and it was soon rising. I'm not too keen on heights, especially in an old wooden basket, and the heat coming from the burner was only just bearable but it was a fabulous sightseeing trip. The landing was a bit hairy to say the least as we were heading into the side of a mountain. I do believe that hot-air balloons can be very tricky to manoeuvre. We ended up landing on a main road and holding up the traffic. Some time later, we read in the UK newspapers that an air balloon had caught fire in China and crashed, killing all on board. Phew. Not many lives left! We also went to an acrobatic show which was very impressive and had Peking duck with Angela, served in the proper way but I prefer it done in the Peking Garden in Southport UK, a Chinese restaurant that my wife and I have been frequenting for it must be forty odd years. I highly recommend the Peking duck banquet.

Me (centre) and two of my technicians on the Great Wall, incl. Dave (left), responsible for the immortal Widow Twanky line.

Shanghai harbour.

GABON

1994

(The 'dense jungle' where Big Foot roams, bugs play tennis and monkeys drive trucks.)

"Monsieur."

"Yes?" I had just landed in Port-Gentil, a small port on the coast of West Africa, on my way to Rabi.

"It is not allowed!"

"What isn't?"

"You cannot bring liquor into the camp."

"Why not? It's not a dry camp – you have a bar there, don't you?"

"Correct, monsieur, but it is not allowed."

'Oh dear,' I thought as I watched the luggage trolley glide past the window heading towards the plane with a bottle of Johnnie Walker inside my suitcase.

The whisky was in one of the new plastic duty-free bottles. I had hand-carried the bottle of Jack Daniel's, with it being glass. After spotting it in my carrier bag, the Frenchman ran to an office to speak to somebody, pointing unmistakably at me. The two of them then strolled over.

"I believe that you have a bottle of spirit with you?"

"Yes I do and another bottle is just about to be loaded onto the plane."

He ran to the telephone. Five minutes later he came bounding over.

"Am I going to get shot at dawn?" I asked.

"You cannot bring any alcohol into a Shell camp."

"So what happens now?"

"I have telephoned the camp security, and they will meet you off the plane in Rabi. They will, of course, confiscate your bottles until you leave."

It would have been nice to have been told about this before leaving. I was here for Christmas and New Year.

We flew on down the coast to Gamba, a small town infested by Shell staff and their families. My boss was there to meet me and to bring me up to speed on the project, which was the commissioning, start-up and testing of the jungle-based gas-gathering stations after the construction phase had been completed. We refuelled and flew off again, this time over dense tropical rainforests which looked like broccoli divided by small winding roads. The usual crowd were there to pick up their mates, back-to-backs swapping their handovers for newspapers like gold. Sandwiched between two ginormous African security officers was a Chinese guy who turned out to be a Frenchman - a Frenchman with his henchmen, who proceeded to do the best Inspector Clouseau impression I've ever seen. He was darting in and out of stacked luggage, head doing a full 360, body-checking people with his eyes.

"Which is your baggage, monsieur?"

"That one."

"Very good."

I was then escorted to the security office by all three in a move so far from surreptitious and cool, even the monkeys were watching from the trees. One would have thought I'd tried to smuggle in a 'berm'.

"In this place, very sorry, it is not allowed, monsieur."

"Yes, I am now au fait with your system." A system that, despite Shell having a bar on the camp serving alcohol, a restaurant serving wine and all contractor camps having their own bars too, didn't allow duty free. I was getting weary and irritated with Chinese Clouseau.

"Can you please give me a receipt?"

"Ah, very good." I found out later that he was called the French Detective.

I was shown to my accommodation which was decent, considering that we were in the middle of an African jungle. After I unpacked and showered, I headed to the main office to meet the team. I was to share an office with two Brians. One, a Marlboro chain-smoking instrument engineer; the other, a Middlesbrough football fan. I sat in the Dutch oil-field manager's office and he brought me up to speed with the history of the oil field. They'd had nothing but problems with the first two gas-compression trains and he didn't want the latest one they were about to commission to end up the same. Shell were loosing a small fortune with the breakdowns.

The gas trains consisted of a gas turbine (basically a jet engine on the ground that drives through a reduction gearbox) and three high-speed rotating gas compressors. I was told to check through the daily logs and see if I could pinpoint any history building up to a breakdown. A job for 'I Haven't a Clue-so,' I thought. Where was he when you needed him?

In the afternoon, I had to take a jungle driving test along dirt roads (there's a speed limit, even in the jungle), doing emergency stops in case an elephant or gorilla stepped out in front of you. Driving along the dirt roads in this part of the jungle was, most of the time, like driving around a maze. The roads were narrow with high walls of trees and bush on either side and just enough room for two vehicles to pass. The scary thing (I was soon to find out) was the logging lorries that would come at you like bats out of hell, only bigger and blinder as they had no radar and could not see around corners. In most parts it was very hilly and they'd be fully laden, so they were 'clogging to get the logging' so to speak, down a hill fast, to enable them to get up the next. Even without the cargo, they were so huge and heavy that they'd be able to take out a herd of elephants, no problem. Also, many of them were from illegal logging companies, so they could flatten you without logging it (sorry). Reporting it would have been pointless anyway. By the time any rescue service got to you in the middle of the jungle maze, you'd have been well on your way to the great forest in the sky (or partying down below). The company medic, great at sticking plasters on fingers and curing tummy aches, would have been out of his depth, on the ocean floor in fact, if, and only if, you could have been either extracted from your now-flattened Land Rover, or shovelled up off the ground.

In the dry season, there'd be red dust everywhere, which made visibility extremely poor: a dangerous cocktail, shaken and scared. The bush was so dense, you'd get completely lost after only a few minutes walking. I suppose that is why it is called dense jungle. This could be really daunting when you were driving alone, miles from anywhere. You had to keep your eyes on the track for obvious reasons and yet you had the inclination to look at the bush, half expecting some creature to come out and greet you. Or eat you.

However, towards the end of my first week I felt quite confident of going off on my own and got hold of a map of the area. As most of the gas-gathering stations were on a circular route, where could I go wrong? And wrong I did go. I somehow got off the road. Instead of going in the circle, I veered to the right onto a disused track. Disused until I came along. It was an old logging route and it kept getting rougher and denser. I couldn't turn round as the track was just wide enough for one vehicle. I stopped to look at the map but couldn't see any reference points, and anyway this track was not on the map. I had no option but to go on until I could find a clearing to turn round. As I was in four-wheel drive and with the weather being hot and clammy, the Land Rover's engine started to overheat. It was also getting dark. There was about an hour or less of daylight and I could smell that distinctive overheating-engine smell. I had no choice but to stop and give it a break, then take a chance and revert back to two-wheel drive in order to take the load off the engine before it seized.

I was beginning to feel anxious at the prospect of having to spend the night out here alone. A little sangfroid was required so I pulled myself together and thought of a survival plan. Most oil companies put you through survival courses, whether it's offshore or the desert, but this was a bloody jungle. In the past, I have been fortunate to have been on several of these survival courses and one in particular was with ex-SAS instructors in the Omani desert. What would they do? I started to put a survival plan together. I had enough water. Good. One of the side windows wouldn't close. Shit. Thoughts of mosquitoes, tsetse flies, spiders, snakes and anything else that would be inclined to feast on me crept in, as they would. Well, I can use part of one of the seat cushions and some wipes that had been left by one of the mechanics after the truck had been serviced, God bless him. What if a gorilla opened one of the doors…? Was my mind wondering? No, hell it wasn't, that was a possibility. Also, a big cat could do the same as the door handles were the old type, a lever that opened the door when it was lowered. A massive cat could, not knowingly and by accident, quite easily do it. I found a rope in the back of the Land Rover - it would do to tie the side doors together by the handles. I could also jam the rear doors. A survival plan was beginning to take shape.

I sat there for what seemed like a lifetime, thinking of all sorts of things that could happen during the night. And also how long would it take a rescue team to find me, if they could. I had worked in the desert before and always carried matches, in case I got stuck in the sand or lost. In the desert, you could set fire to the spare wheel and a spotter plane would be able to see the smoke for miles. I was in the middle of a jungle on a disused and overgrown track. There was no way I was starting a fire. The golden rule in the desert was to never leave your vehicle as you could be walking round in circles not far from it, using up your energy and water. This actually has happened, where a guy left his truck and walked for hours under a deadly sun and ended up back at the truck. He was eventually found dead, holding a letter he had written to his wife.

I made a decision to restart the slightly cooled-down overheated engine and just carry on. Even though I had the tendency to go fast, I resisted, not wanting to seize the engine. After about half an hour, to my great relief, I came to the main road. The euphoria and relief was stupendous. Saved from the claws and jaws of death. Fear and elation, what feelings.

"Where did you get to?" was the question when I eventually got back.

"Oh, just looking at some of the stations."

"Did you find your way okay?"

"No problem." Said with a hint of a tint on my face.

The bar at the Shell base camp looked rather exotic from the outside with its grass roof. It was at least forty foot in diameter. I went in for a drink before dinner; it was dark inside and somewhat bare and there weren't any glass windows, only wire-mesh screens to keep out the jungle nasties. It had three electric ceiling fans that rotated at least ten times per minute, so five minutes after you'd showered and put on clean clothes, you were a clammy mess. The bar was small and L-shaped and seemed like an afterthought. In the corner were two old refrigerators. Several ultraviolet bug-catchers were dotted around the place and I was soon to find out how vital they were. You'd be stood at the bar having a drink and instead of hearing a small cracking noise, like you'd hear in a fish-and-chip shop when a fly gets zapped, there'd be a loud bang, with dust and body parts of some huge bug cascading everywhere. You soon got used to covering your drink and your neck double quick as soon as you heard an explosion. If you were sitting down in one of the armchairs, a lizard or a spider would run over your lap and your drink would end up on the ceiling.

The refrigerators gave out about the same power as a small torch, minus a bulb. It was embarrassing when we invited people from another camp to a function, as we'd always run out of cold beer. The contractors' bar put Shell's bar to shame as it was far superior with huge industrialised refrigerators and air conditioning. Saturday nights were an open invitation to the Shell crowd. You could drink as much as you liked and a slap-up meal was always on, usually a curry with all the trimmings and wine served at the tables, all for a fiver! We, working for one of the world's biggest oil companies, felt like the poor relation. Oh, and they had Walkers crisps at the bar too. Fancy! After the meal, we'd return to the bar for a quiz and a lot of banter. Which was great fun, unless you lost sense of time…

One night in particular after the quiz, I got talking to a fellow Scouser and Everton supporter, who happened to be the uncle of a mate I trained with back in the sixties. I was so engrossed in the conversation that I didn't notice my colleagues leave. I thought it rather early for a Saturday night. Our bar closed at nine, so what had they all dashed back for? To sit in their rooms?

"Have another drink," said Scouse. "You can stay here, the camp boss will give you a room."

We continued talking and reminiscing about his nephew and what he used to get up to. He was wild then but seemed even wilder now and had somehow got himself fixed up with a job on a billionaire's yacht, based in Monaco, the 'dog bog' of the French Riviera, on a high salary. Then tragedy had struck, as one of his daughters had died in tragic circumstances and he was now divorced. It was getting late as I took all this in. I was tired, felt subdued and a little inebriated. The camp boss had gone to bed and I didn't want to wake him up to get keys to a room as I somehow felt that he would not be pleased at all so I decided to walk the mile to our camp. Through the dense jungle. Bravado strikes again, with a hint of Dutch courage.

The night air was cool and there was a full moon, no problem. I walked or rather zigzagged along the dirt track and came to the barrier which was down and locked. I didn't want to go around as it meant a detour into the bush, so made the decision to 'limbo' under the bar. I tripped, fell forward and caught my chin on the ground, which started to bleed. The jungle now had the aroma of fresh blood and it was night-time. Feeding time. Believe me, this was an instant cure for drunkenness, as the realisation of my predicament started to dawn on me. I carried on walking. The moon had now gone behind a cloud so I could barely see where I was going. In the jungle, day turns into night like switching off a light and from hardly any noise to mayhem, as all the creatures come out to dine. They are killing or being killed, eating or being eaten. I had reached about halfway, when I saw something move on the track ahead of me. I stopped, as I couldn't make out if whatever-it-was was coming towards me or going in the opposite direction.

The background noise had now reached a crescendo and was rather scary to say the least. My heart was pounding. What was it? My eyes were now adjusted to the darkness and I could make out what it was. Elephants. Oh shit. There were two adults and their baby, a very dangerous situation to be in. They had now stopped, their ears splayed. Obviously, they had picked up my scent - lager and blood! The waiting was agony as the noise progressed from being scary to damn terrifying. I could visualise big cats, gorillas, snakes and all sorts of horrid things coming to feast on me. The elephants started to move again and disappeared into the thick bush.

I stealthily carried on but had to get past where they had gone off the track. Were they waiting? I arrived at the spot. My heart rate must have been like an astronaut's, waiting to blast off, as acute anxiety crept in. I did a Usain Bolt with his hair on fire, back to my room, just remembering to open the door before I burst through it.

Elephants are fantastic beasts but not to be messed with. I used to see them quite a lot and had several encounters with them. These elephants were a brown colour and well-camouflaged in the jungle. I was jogging one day along the track when I nearly bumped into one that was having a nibble. I did a quick about-turn. Problem is, they can outrun you, as big as they are. Another time jogging along the airstrip (when it was closed, that is) with a mate called Greavsie, discussing something or another, we heard this almighty bellow. We had almost run into a herd of them. It was just like one of those Laurel and Hardy movies as we stuttered and sputtered, bumping into each other trying to run for it. Fortunately for us, I think the elephants couldn't run for laughing.

Another time, we found a baby elephant that had been split up from the herd. The poor thing was about the size of a big dog and I caught this French bloke trying to pour milk down its trunk.
"I do not think the elephant likes the milk!"
"Put the milk in a bucket, you wanker."
I don't think he understood and neither did the elephant. The only thing we could do, apart from saving it from the demented Frenchman, was get the local security guards to load it on a truck and try to find the herd, or a herd. It was a regular occurrence to open your door in the morning and find an elephant turd on your doorstep instead of a pinta. The camp didn't have any fencing to keep out any beasties. One night I was kept awake by a frog outside my room. I flashed my torch through the window and the little sod stopped, then

as soon as I got back into bed it started again. There was no end to this – it went on and on. I took a chance and went outside to locate and chase it. I got back into bed but as soon as my head touched the pillow it was at it again. A thought came to me. I still had some of those little yellow foam earplugs the ones you get on airplanes. I put them in. Great. Just as I was dozing off, I felt something run down my chest. I was out of my bed like a shot, expecting to find a spider or something just as horrid. It was a bloody earplug.

Tennis is, for me, a great game to play and Shell being Shell had a tennis court here, in the jungle, of all places. At different times of the year, we would get plagued with different types of pests. One in particular was the rhino beetle, aka The World's Worst Flyer. They are the most non-aerodynamic flyers that fly and are the size of a mouse, with rhino tusks on their bonnets. They just crash-land most of the time, mainly into walls, or you. One day, during a game of tennis, I was just about to return the ball, when one flew right at me. Thinking that it was the ball, I whacked it back over the net. I didn't get the point and it was a great shot too. You know how some of the prima donnas of the tennis world would go bananas if somebody crunches a crisp or coughs at Wimbledon? Well, they should try the jungle. As you were playing, you'd hear an almighty bellow or roar, whilst being eaten alive by mosquitoes. On leaving the court one night, I almost trod on a snake, a green mamba that was about to devour a frog that it had in its mouth. Feeling sorry for the frog, I picked up a stone and threw it at the snake. With that, the snake let go of the frog and the frog, which was yellow and poisonous, leapt at me, hitting me on the chest. I brushed it off, as you do, and whilst doing that, the snake, also poisonous, ran across my foot. Not exactly Centre Court Wimbledon but nearly game, set and match.

"You were lucky last night." I was chatting to my Scouse friend, when Gerhard the Austrian came stumbling over. He was part of our tennis foursome. I teamed up with Pete the Taff and we used to thrash him and his partner.

"What do you mean, lucky?" Of course I didn't mention my lucky escape from the fangs of hell.

"You were crap." The duel was starting.

"I bet I could win you at the singles," he said.

"Bollocks." Both of us by now were well oiled.

"Fifty quid says I will win the match."

"Bloody cheapskate, make it a hundred." (Golden rule No 1: Never arrange a duel or a bet when pissed. Golden rule No 2: Never play the next day with a hangover, especially in the jungle with the temperature over 100 degrees.) The first set was a piece of cake, I won no problem. Unbeknown to me, he'd been drinking this rehydration drink all morning and, with him being younger too, it was drinks all round, on me.

"Are you free?" said one of Shell's managers. I was thinking of John Inman.

"Yes I'm free, what can I do for you?"

"There's a disused guard hut in the bush and I'm thinking of bringing it back here and using it as a rugby clubhouse, next to the ground." What he was really saying: 'I would like you to arrange for it to be transported back here and transform it into a clubhouse.' "We can drive out into the bush this afternoon see what you think." The thing that you tend to find with big oil-company staff is: they get an idea and get you to execute it. If all goes well, they receive the accolades. If it doesn't, you get fired - type of thing. It was December and, as I wasn't Father Christmas and snowed under with work, we drove out into the bush and sure enough the old guard hut was there. "OK, then I will leave it in your hands."

"Oh, thanks, super." So I was left with project clubhouse for the time being, until my equipment arrived. At least it gave me something to do. First, I had to arrange for a truck with a pipe tied horizontally to the rear, the same width as the guard hut, so that we could gauge if we'd be able to negotiate round all the trees and narrow bends in the track. We achieved this without a hitch. The next problem was to arrange for a crane to lift it onto the lorry. To get a crane was a hurdle; to arrange for a lorry was Becher's Brook at the Grand National, Aintree. We also required a concrete base for the clubhouse to sit on. This is where the wheeling and dealing started and I had to do a Bernie Ecclestone with the local contractors. All this because the Shell field manager wouldn't put any money in the kitty for our project. The contractors knew if they upset one of the chinless wonders from Shell they may never get another contract, so they agreed with their arms up their back and the job was done. The contractors had been Bernie'd and the clubhouse eagle had landed, in the words of Neil Armstrong landing on the moon, a feat almost as unlikely.

The hut had to be debugged, of course, as all sorts of horrible things had been nesting in it. In doing this, the bloody ceiling caved in and required replacing. I was running out of favours from the contractors, who were getting a bit edgy with us using their labour, time and equipment. Pete the Taff and I managed to obtain a nice piece of hardwood as a bar top and a large refrigerator. Another contractor chipped in with a stereo system. The place was painted and Greavsie sorted the electrics and water supply out. We put netting around the veranda to keep out the mosquitoes and other critters, then seconded some tables and chairs and a large, spare, unbreakable, glass door from the squash court. I said that it would be ready for New Year's Day and it almost was but it didn't stop us from having the Grand Opening. I had managed to obtain two bottles of cheap champagne from the camp boss. The tape was to be cut by the Dutch oil-field manager who showed up with his camera. "Mm, very good."

I pointed at two hairy-arsed drillers throwing a rugby ball to each other.

"Bet that will go down well in your annual HSE report - how you keep the men fit?"

The champagne was opened, clubhouse toasted, speeches made, photos taken. Great. Then after the champers, it was dampers.

"I hope that you are not going to have alcohol in here?" said the manager.

"Bastard" said I! End of clubhouse. Opened and closed quicker than you could order a pint. What a complete waste of time and effort. I sometimes wonder what planet these people come from. They already had a bar on the camp. Was he expecting? A rugby club without beer?!

Bang, bang, bang! 'What the..?' It was three o'clock in the morning and somebody was knocking hell out of my door and it was no elephant. I looked out of the window. It was three of Clouseau's security thugs, dressed in ominous black. I gingerly opened the door.

"Mister, mister, come quick."

"What's the problem?"

"Come, mister, this man is sick." I quickly dressed and followed them. Several security guards were outside Marlboro Brian's room.

'Oh shit.' I dashed in to find Brian gasping for breath. He'd been warned off the cigarettes but had carried on smoking and was in a bad way. Although it was painfully obvious, I asked, "What's wrong, mate?" In-between gasps, he blurted out that he couldn't breathe.

"Where's the medic?"

"He come, mister, he come."

"I'm a goner, mate," Brian said in-between gasps.

"No you're not, Brian, you're a bloody fool." He needed oxygen urgently. I hoped and prayed that he wouldn't expire before the medic came with the oxygen, as I didn't fancy giving him the kiss of life with all that puffing and panting. The medic was taking ages and Brian was very nearly on his way to the great Marlboro factory in the sky.

"Where the fuck is the medic?"

"He come, mister, he come." My despondency was creeping in. "This him now!" A man shot through the door clanging the oxygen bottles.

"How bad is he?" said the French medic.

"Well, if you don't hurry up with the oxygen, he will soon not require it," I said, out of earshot of Brian. The medic then telephoned the Shell doctor in Port-Gentil for, believe it or not, instructions. This really made one feel confident in our medical facilities. I looked down at Brian who reminded me of a puffer fish, with his cheeks inflating and deflating. He was making a dreadful sound. Instructions were given and taken, the medic gave him oxygen, then gave him an injection and put him on a drip. For an agonising period it was touch-and-go. He had to be medevacked to a hospital but it would take the plane two hours to get here in daylight. I made a mental note not to have an accident until I got home. A thought struck me: I have never had my appendix out.

Minutes seemed like hours as we waited in hope that he would stabilise. In the commotion the guards had left the door wide open and now the place was full of mosquitoes, which were not only attacking Brian but me as well. I covered Brian as best I could. For the next hour I sat next to him fending off the evil little sods while the medic had a break. At last he stabilised and seemed to make a miraculous recovery, sat up and asked for a drink. I made tea and brought it to him. I couldn't believe what he said next!

"Pass us the cigarettes before the medic comes back." That is the honest truth. I had been up all night assisting the medic, trying to save his life, fending off mosquitoes and organising things.

"You can fuck off, Brian." I had visions of him inhaling oxygen in-between lighting a Marlboro and, as I didn't wish to be blown up due to the mixture of oxygen and flame, I flushed his cigarettes and matches down the toilet.

"You bastard!"

"Cheers, Brian."

He was flown out the next day and I have never heard of, or seen him, since. I wonder if he did go to the great Marlboro factory in the sky. Silly sod. It was now 07:30 and I was completely shattered. The oil-field manager appeared.

"How is he?"

"He'll live."

"Can you make out a report?" Just what I needed after being up all night. That's life, hey?!

"Fancy a drive into the bush?" asked Kevin.

"Yes, why not?" It was Sunday afternoon. Kevin was the mechanic from the drilling company and had been in the jungle for five years. I had given him the nickname 'Daktari'. He certainly knew his way around, as you would after spending five years here. After lunch, we set out and drove for several miles – not many signs of life apart from a few monkeys and a solitary elephant. He stopped the Land Rover. "Let's take a walk," said Daktari.

"You sure it's safe?" said Chickenari.

"Oh sure, been here loads of times." It was very spooky, not a sound, which somehow made it feel twice as perilous. I constantly kept looking over my shoulder half expecting – well, expecting – something nasty to appear. We were unarmed and extremely not dangerous, which didn't help the nerves one iota. I kept thinking what we'd do if something attacked us - apart from having exceptional bodily functions and running like hell, which was about all that we could do. We carried on. Dak was a few metres ahead of me, looking far too confident, the smarmy bastard. I picked up a stick and threw it into the bush right next to him – boy, did he jump.

"What the fuck was that? A big cat or gorilla?" I said, immediately scaring myself in the process. "Do you think we'd better head back?" I used to think that I was a phlegmatic type of person but the word 'nemesis' comes in somewhere here.

"We're nearly there, just over this hill." Shit, why do I always let people talk me into these situations? We reached a sandy clearing by a lake - ideal crocodile terrain.

"What do you think?"

"Oh super, Kevin, really great, now let's get the hell out of here before we are turned into croc chocs." The bastard saw my acute anxiety and was plotting his dastardly revenge. I was now ahead dictating the speed, which was more like an Eastern European speedwalker.

"Ken, Ken, look at this." I skidded to a halt, leaving tyre marks on the ground.

"What?"

"Fresh paw marks in the sand." I looked, I froze. They were about the size of Muhammad Ali's fists with gloves on. Even minus my jungle tracker's ability, I knew that they were fresh. I detected a smirk on his face. The bastard had brought a wooden paw with him, to trick jungle rookies like yours truly. We trotted off, laughing our heads off, with a hint of fear in our laughs. We climbed back into the Land Rover and headed off. Our smiles soon diminished.

"Tsetse flies."

"Oh shit."

"What now?" Daktari was turning the truck round in the sand. We had guests. The truck was full of them and we daren't stop in case we got bogged down. A tsetse fly is a sure cure for insomnia. If bitten, you are likely to catch sleeping sickness, which as you can imagine is not entirely a good thing to catch and slightly inconvenient too. I was getting rather testy with a tsetse, a real jungle nasty in a similar league as the mini vampires (mosquitoes). [As I write this, I am in Angola, West Africa, previously known as the white man's grave. I have taken my prophylactic tablets, I have my air conditioning on quite high, my door is always closed - well, at least when I am in my room - but one cannot rely on the house boys keeping it shut when they clean your room. At night I spray my face, ears and head with jungle formula insect repellent and wrap my bedsheet around me and they will still get at you!] In one of the quickest U-turns in the history of U-turns, even quicker than an F1 racing car, Daktari gunned the Lambo Rover with his size 10. [Just killed a mosquito in my room, on my desk, see what I mean? And it is bloody freezing in here too]. We opened all the doors and windows to blow the sods out.

Daktari Kev told me a story (how true it is, I cannot tell but I can well imagine) about some Americans that had been working here. Kevin had taken them to the same spot. The day before, he'd taken a very big pair of wooden feet made, with straps on and walked around the sand, as you do. Of course these gullible Americans swallowed, not only the bait, but the rod and the reel too, followed by the maggot box and chair as a dessert. They were full of it, the tale that is. They immediately assumed that Big Foot was residing there, phoned the National Geographic Society and had the area cordoned off. I know that I was gullible with the paw marks but would I have done the same? No, surely not. The day before the Americans left, their British colleagues held a going-away party for them. And guess what they saw hanging over the bar? Yes, it was our Kev's wooden feet. Good job that the National Geos or David Attenborough didn't come out with a film crew.

It was Sunday afternoon and Taff and I drove over to the contractors' camp. On the gate was a huge security guard, a bit like Idi Amin.
"That's Chopper," said Taff.
"Why is he called Chopper?"
"He went home on leave last year and found out that his missus had been having carnal knowledge with the local milkman, got a bit suspicious of getting free milk, gold top, in fact, and chopped her head off so now he can't return, as her family were a bit fed up with that."
What a way to settle the score.

"What happened to Ernie, the fastest milkman in Africa, then?"

"Nobody knows, he's never been found."

We got to the bar and ordered two beers. It had been a busy afternoon judging by the empty glasses and bottles. Suddenly a monkey ran along the bar top. It was one of the monkeys that had been rescued and lived in a cage at the rear of the camp. Their mother and father had been killed by local tribesmen for bush meat so the kids had been adopted by the guys at the camp. Every weekend they were allowed out of the cage to roam the camp. The bloody thing had picked up a bar towel and was flicking it like a whip at the local barman and amazingly it did not knock anything over. Monkeyana Jones nicked a packet of cheese-and-onion crisps, ran out of the bar and straight up onto the roof followed by an irate barman. You had to be careful when you went to see them in their cage because, as you approached, they would slyly pick up some stones and, when you got closer, they'd let you have it. Hilarious, until you got hit. They used to do this a lot, especially if they had a hangover. Oh yes, some bright spark had been giving them copious amounts of beer. Alcoholic monkeys. Was this the reason for their visits to the bar and their rage at the barman who refused to serve them? It could only happen in Africa.

Christmas Day. We had to turn up in the morning, then a quick shower and change of clothes. Pre-Christmas-dinner drinks. I was looking forward to tucking into turkey, roast potatoes, etc. but I couldn't believe what was on the menu: crocodile stew. I looked at it being served; this was a delicacy to the locals. I just pictured one of David Attenborough's Planet Earth documentaries where a crocodile would surreptitiously glide over to everybody's fall guy, the wildebeests. Whoosh. One beast would be gobbled up by another beast. And life goes on – well, at least for the crocodile. I had to try it, just a bit, and yes, it tasted like chicken. Just like frogs' legs. I just didn't fancy sucking their toes. Shell's choice of Christmas dinner for a load of Europeans.

We were all invited for drinks and goodies at the contractors' camp in the evening.
"Is everybody here?" We were about to return to our camp. "Where's Taff?" I knew where he'd be. I went straight to the African bar. He was dancing with all the locals who were all in a stupefied state, including Taff. God only knew what they'd been smoking. How do I get the silly sod out of here and back into his mind again? The atmosphere was very intense, sort of inter alia, voodooism unchained. Rather than risk intensifying the intense by breaking up the party (as if), I left Peter to peter himself out. He was obviously used to all this as he watched rugby matches at Cardiff Arms Park, unlike us civilised gentlemen at football grounds in England. I waited until the delirium had died down, or I should say, 'when he was basically effed.' He was like a sweaty rag doll.
"What was all that about, you big daft Welsh c*nt?"
"Don't know, boyo, but it was good, eh?"
"Oh sure, Pete, enjoyed it tremendously, now let's get the hell out of here before they start up again."

I returned from post-Christmas leave, armed with a gorilla's mask, which I used to good effect. This particular day, I was driving along the track to one of our stations, when I spotted one of our Land Rovers that had parked up. I donned the mask, a dark-blue North Sea rig quilted jacket and black leather gloves, feeling just like an extra from 'Planet of the Apes', then pulled up alongside a little Indian surveyor setting up his theodolite. He looked at me in sheer terror. I growled and drove off, weaving all over the track and waving my arms. His face was a picture. The gorilla turned up again one night outside the Shell bar. I crept up and banged at the window, or I should say, 'scratched the wire mosquito net', and growled, running off as I heard their drinks smashing to the floor.

The HSE department spent ages trying to find the culprit, never did. Also in my armoury was a rubber snake. A bit of black cotton round that and when one of the local secretaries came into our office, the snake would shoot across the floor and they'd run out screaming blue murder. For good reason - the locals hate snakes. I'm not too fond of them either.

Our cleaning boys always left the door open when they were doing your room. I was so fed up telling them to keep it shut. Of course all the flies and other nasty things would get in. One day I was driving past and saw my room door wide open. The two cleaners were still inside. I stopped, got out the rubber snake, threw it in the room and closed the door. I have never seen anybody move so fast. They never kept the door open again. As I was retrieving the snake I noticed that Middlesbrough Brian's door was open. He was known to be a mouthy little prick so I put the snake in his bed. Funny, never heard a thing about it. But he was very quiet for a while after.

Working for major oil companies can be very strange. They'd do daft things like have an al fresco buffet dinner. Now imagine eating chicken and steaks etc, outside in the middle of a jungle in the evening. Before you could eat anything, it would be eaten for you and of course you would be turned into the Elephant Man by the flying bugs of the night. It was an entomologist's dream; they could have held a world conference there. Every week we would have a camp fire drill, which is fair dinkum, but what I couldn't understand was, why they had lifebelts outside the accommodation. I mean, we were in the middle of a jungle, miles from anywhere - the only water was in the fire buckets. Maybe they thought somebody could drown in one of those. Of course we have to have the elf and safety executive but they can, and do, go over the top. Mr HSE wasn't at all pleased when I showed up at a drill wearing one.

I was once working at a Texan shipyard, where it was compulsory to wear a hard hat inside your car on site. Now imagine that your car has been outside the office in the Texan sunshine for hours and you have to drive to the site for something - you couldn't leave any windows open because of the dust. As you opened the car door the heat would hit you smack in the face like putting your face in an oven on a Christmas Day. Now it's bad enough getting into a car that had been outside in the sun in the UK, that is when it shines, but the middle of a Texas heatwave, wearing a hard hat! I've been on a project where you have to wear a full safety harness when going onboard a vessel, even though you are not going to climb or work at any height. This is on top of a flame-proof coverall with long sleeves which is hot and heavy, plus safety boots, gloves, safety glasses and a hard hat that has ear protectors attached. In temperatures of 100 degrees. Health and safety!

On a Sunday, Shell would allow us to finish half day, so that we could show off our sporting prowess. We had to go through with this exercise or the privilege would have been taken off us for sure. Most of the guys liked to show their dexterity through arm wrestling, with a glass preferably. However, we organised cricket matches on the airstrip and I must admit I enjoyed it. We played against other contractors and a few niggles and grudges would come to the surface. There was a large planning engineer who I could not get on with. Every time I went to his office, out would come the diatribe. To my knowledge, I had never done or said anything bad about the 'fat bastard'. He was my bête noire. This chubby chap rather fancied himself as a bowler and would start the bowling. I was opening batsman. He was already wound up so to wind him up more, I would stand like an American baseball player, instead of guarding my wicket – this really pissed him off. (Note, this was after his diatribe.) He would come charging like a white rhino, steam coming from his nostrils, and almost throw the ball at me like an American baseball pitcher. I used to play for the school cricket team, so I wasn't too bad, especially against rhinoceroses. I walloped the first ball for a six. I kind of sensed that he would like to kill me. It was probably six of one and half a dozen of the other but I could not fathom him out, or what I had done. You know the type? Revenge is sweet, so they say. My bête noir managed to get his revenge when we had a football match. I was in goal and he was the referee. Nuff said. One of our players passed the ball back to me and I picked it up. Having not played football for years I wasn't au fait with the new rule, where a goalkeeper wasn't allowed to pick up the ball if another player from the same team passed it to him. He immediately blew his whistle. "Penalty!" he shouted, as he jumped up and down with glee. I think he actually gave the goal before the ball was even kicked. I stood no chance. He actually ran round with his shirt over his head like a Premier League striker who had just scored to win his team the FA Cup.

We were playing an African team and our cricket umpire was called George, an ex-naval officer. He would appear in all his officer whites: shorts, shirt, long socks, white shoes and white beard. Officer to the core with his ensign still stuck up his bottom too. He would stand behind the wicket, arms behind his back, rocking to and fro like an admiral of the fleet. George spoke so far back that he would sometimes fall over. If anybody whosoever had the nerve to be vociferous about any of his decisions they would be immediately keel-hauled. One of George's degrees was in philosophy and the jungle was not quite the place to use it, especially with a vociferous Scouse git - no, not me this time but a fellow Scouser, known as 'Our Ray'.

"Ah-rey! That wasn't a foul, you blind c*nt!"

That kind of expletive just did not go down to well with an ex-naval officer, come admiral of the fleet, D.S.O with a degree in philosophy and another in speaking posh. The Scouse git was soon turned into stone by one of his authoritarian glares.

One evening, I had been invited to Taff's for a drink, along with George and a Glaswegian driller. We were sat, having a drink and a chat and, as the night progressed, George stood up, rotated Taff's wall spotlight towards him and went into a rendition of Felicia Hemans' poem, 'Casabianca', better known as 'The boy stood on the burning deck', in his mellifluous officer's voice. To say that 'one could hear a pin drop on a shag-pile carpet a mile away' would not have been overstated. Jock's jaw dropped sufficiently enough to amputate some of his toes, followed by: "What the feck you on aboot, Jimmy, you big daft coont? Sit yersel doon and ave a swally, the noo!"

Another one of Gorgeous George's favourites was: "Anybody got any mons?" Mons meaning money. We would be in the bar. "I haven't got any mons on me, be a good chap and I'll sort it out with you in the morning."

Well, what can you say to that? "Thank you, George, it will be an honour." When really you wanted to say was: "Sod off, you cadging old fart!" He was one Machiavellian character. He used to wear these ridiculously tight jeans; I'm sure he had to grease his legs to get them on. He was totally out of tune with any fashion – or I should say, any era – but it obviously didn't bother him in the slightest. He reminded me of the comedian the late Max Wall, who used to do a sand dance on stage wearing black tights. George was pure burlesque.

We were all invited one weekend – or I should say 'told' – to attend a party that Shell were throwing at their camp in Gamba. Not to turn up would have been a slap in the face for them and, of course, we didn't wish to bite the hand that fed us. We travelled for at least three hours through the jungle, following some louts, who every so often would toss empty beer cans out of their truck. I suppose one could call them 'tossers'. We reached the lake and clambered aboard a launch to take us the remainder of the way. George met us at the other end, along with his unswerving and lifelong lackey, who was George's junior officer in the Royal Navy. George, of course, was wearing his customary spray-on jeans, now with pre-George Michael holes in them. Fortunately it was a fancy-dress party but he didn't give a toss anyway I'm sure, as I cannot imagine anybody wearing those jeans and worrying about what people thought of them. I often wish that I could be like that myself. The last I heard of George, he was living with a female welder with tattoos. As Shakespeare kinda said: "By George!"

Waiting for the next boat to jungle head office. A colleague and myself (left) in our Royal Dutch Shell colours.

South Africa 1996

(Tusks, Jaws and Davy Jones's Locker!)

Our FPSO docked in Richards Bay. (Floating, production, storage and offloading unit is a floating vessel used by the offshore oil and gas industry for the production and processing of hydrocarbons, and for the storage of oil.) After a night's stay, we had a crew change before flying to Johannesburg for the connection to the UK. It had been an interesting voyage that almost ended in disaster on the first night at sea. As we were passing North Korea, out of the mist came a huge Chinese iron-ore carrier, fully laden and heading straight at us at full speed. We were also a huge ship, new and bigger than a supertanker and one with huge empty oil tanks as big as cathedrals and being towed by two large ocean-going tug boats whose captains blasted their horns and directed their full-beam searchlights at the Chinese ship. We were told the next day that nobody was on the bridge of their ship and only the quick thinking of the tug-boat captains averted a tragedy. They had slowed down, which slackened the huge wire-towing lines to allow the ore carrier to pass in front of us and over the wires. I was in bed at the time, watching a movie on my laptop and had no idea what was going on and didn't hear anything due to wearing earphones - rather stupid of me and taught me a lesson! If the ore carrier had hit us, even with our huge tanks being empty, we would have sank like a stone in minutes. I would have got quite a shock as the seawater covered my laptop. I do believe that their captain was spoken to when they reached their destination in Singapore. So I missed the Grim Reaper and Davy Jones's locker and try not to count my lives left, if any!

Good things on the voyage were seeing lots of whales going on their annual feeding vacation, the wonderful sunsets and the incredible sight of Table mountain. We changed crew in the hotel in Johannesburg for the next team to carry on to Angola where the FPSO was to be stationed over the oil well. It would take at least two weeks before the ship would arrive, so instead of flying home, I took the opportunity to take a vacation here as I'd never been to South Africa before and telephoned my wife to get her to fly out. She didn't need much persuading: I think that she was on the next flight before she put the phone down.

South Africa has such a lot to offer the tourist, so we stayed at the hotel and decided to get advice from the concierge the next day. They recommended that we take a tour of Soweto as it was quite near our hotel. We first went to see the Apartheid Museum and then onto Nelson Mandela's house, which is now a museum, passing Archbishop Desmond Tutu's and Winnie Mandela's houses, which were close by. It seemed odd seeing a pair of his shoes by the bed where the great man slept. Our guide advised us to look around one of Soweto's shanty towns. Although the people were friendly, it felt strange and a bit uncomfortable but well worth seeing. The next day, we hired a car and drove to Durban, where friends of ours let us stay in their holiday home. It was a very easy and pleasant drive. We stayed that night in a wonderful hotel in the Drakensberg mountains and next day hired horses for a ride around them. We also rode to Rorke's Drift where the infamous Zulu battle took place; I felt a bit uneasy as one didn't know what could be lurking around in the bush! The following day we took a tour of a Zulu village and were invited into one of their homes which just happened to belong to the witch doctor. I ended up sitting next to him as he preceded to carry out his witchcraft, juggling bones on the floor. Old witchy-coo offered us a drink of 'I do not know'. I declined the offer as everybody else had drank from the same cup and I didn't want to end up in the pot. Or on it. This was followed by a spectacular Zulu dance. It was quite a remarkable visit and truly wonderful experience.

We continued our drive and soon arrived in Durban. We had to get directions at a travel agents to find our friend's house, which was in a gated community. Apart from keeping thieves out, it also kept monkeys out as there were many trying to get in (bit different from Liverpool, monkey-wise I mean!). The estate agent tried to sell us a house overlooking the ocean; it was absolutely gorgeous but a bit out of my reach, a lot actually. It must be great to be rich and buy what you like but to have a look around was enough, so you keep on buying lottery tickets and throwing your money away… We promised a South African friend of ours that we would bring some footballs out with us for his charity. These were to give to some children at Spion Kop (the original one and not Liverpool FC's). We drove along and came across some children playing. They came rushing over to us and we gave them the balls and a pump.

When we were in the travel agency, we booked a stay at a private game reserve called Tanda. The accommodation were small round huts but beautifully furnished, with an outside plunge pool where you could sit while having cocktails and watching the wildlife and scenery. The first night we went to the restaurant but unbeknown to us we should have called for an armed escort, as we were later informed by the staff that lions had a tendency of patrolling the grounds. We breathed a large sigh of relief and quickly ordered some drinks. Imagine going for a meal and ending up on a lion's plate! The first bottle went down rather quicker than usual! Next day, we went out on safari in an open Land Rover with an armed escort and came across a pride of lions with bloodstained mouths. Their victim had been a zebra as we spotted a baby zebra running around in a panic looking for its mum. After spotting two cheetahs, also post-dinner, our driver stupidly drove up to some very large elephants. The largest one happened to take a disliking to our intrusion and started to charge us, which was not a wonderful experience but hey, who would like to be gawped at when having a meal? It is their country. We quickly reversed out of danger...

Having a shower in the morning was something else as you would get some of the wildlife walking past, not bothered at all – at least I hoped that they weren't. The shower was refreshing but your heart did thump.
 "Oh what happened to him?"
 "He was eaten when he was having a shower."
 These things don't happen back home whilst holidaying in Butlin's or in a caravan in North Wales. But this is Africa where it can and who wants to go to Butlin's or North Wales anyway?

That night, after surviving being the main course at a lion's dinner and being a battering ram target for an angry elephant, we went on a night-time barbecue al-fresco style. Obviously it started to rain, as it does at barbecues, but we were covered up, so not a big problem. The rain stopped and the fires were soon lit but I did feel a wee bit uneasy as to what could be crawling in the undergrowth and what was on the menu, so long as it wasn't me. We had only booked a week at Tanda private reserve, so we had to move on. We booked into a government reserve which turned out not to be as good, but with no other choice we stayed. Our first safari happened to be a night one. I thought that it would at least be different, which it was and we came across many wild dogs and eyes shining everywhere in the glow of our headlights. Then our driver spotted two rhinoceros, one being a male who I assumed was feeling rather frisky and was chatting up a female. Our obviously inexperienced driver shone a huge beam at him. Well, the boisterous rhino was not too pleased about that, as you would expect as he had probably taken her out for a meal and a champagne dinner and the driver spoiled his dastardly plans. It came at us like an express train, but faster and breathing more steam. I said to my wife as casually as I could pretend, as my eyes were bulging and almost popping out, "I think we had better move to the other side of the vehicle." Not that it would have made much difference. Our idiot driver hurled his searchlight at the angry charging train; it hit the poor beast and luckily it stopped in its tracks. There was now glass all over the place from his now-broken light, ready for some innocent animals to injure themselves on, but we survived to be prey for another day. The endangered species also survived to charge another day before he made us extinct.

Then it was on to Cape Town. What a fabulous place. We had already booked into a B&B, which was just perfect, not too far from the beach, various shops and restaurants. The weather was perfect too and we had plenty of walks around the harbour. I did a stupid thing and drove through a black township, which I was told by our B&B proprietor was a 'no go' thing to do but I must admit that we never had any problems at all. We were also warned about the likelihood of being carjacked and that when you came to stop behind another vehicle, you should leave a space behind them so that if somebody was trying a carjack or rob you, you could at least drive out of danger and not be trapped in-between two vehicles. We never found ourselves in that situation but I believe that it does go on. Life is just one big learning curve, isn't it?

At the harbour, we went into a tourist shop and I spotted a big poster with the movie star Nicolas Cage inside a cage on a dive with great white sharks. I had to have a go, so we booked this for the next day. The driver picked us up at four in the morning as it was a long drive. When we arrived we went in for breakfast as we were on the road too early to have had any at our B&B. At breakfast at the shark restaurant, we were subjected to a wall of 52-inch televisions showing huge great whites hurling themselves out of the water with a large seal in their jaw. Just the kind of thing that you want to see before you go into the cage. I did not eat too much at breakfast. I think they show this repeatedly on purpose so that you don't eat much and save them money on food and washing-up.

We got into the RIB (rubber inflatable boat) and set off for the main boat which I was relieved to see as I didn't quite fancy our chances in a rubber boat after seeing Jaws, the movie. After a bouncy, choppy ride, the main boat seemed a decent size. The cage seemed a wee bit on the small size to me but what the hell, no turning back now. We put on our wetsuits. I was expecting an air bottle or a snorkel but was only given a mask as our captain said we'd scare the sharks away with the noise of our breathing. Then the crew started to throw chum in the water, blood and guts. The cage was lowered down the side of the boat and he asked for a volunteer to go first. Nobody volunteered as expected but James Bond stuck his hand up, didn't he? I thought it only right as I am a qualified scuba diver (but unqualified shark bait). This seemed a piece of cake - well, getting into the water and the cage was, I had forgotten about the killing machines that were going to come and say: "Hello, food."

I climbed into the cage and was told to go to the end, where I had the side as well as the front to look out of. The cage bars had a gap of about a foot and I had the side to contend with as well. I knew that a great white shark couldn't get through that gap but if you did see the film Jaws, remember the huge hole it made in the cage? Exactly, you get my point! The captain told us that if anybody put their hands out to try to stroke one of the sharks, he would pull them out, probably in pieces. As if you would! He sure didn't have to worry about me. The only way to view the sharks was by holding your breath. As we had masks on it wasn't too bad and, after a few minutes, I took a deep breath and went down and immediately spotted a shape coming towards us about twenty to thirty metres from the boat and imagined the music from the movie Jaws in my head. I shot up to the surface to inform everybody on deck that a shark was approaching. It whizzed passed the cage by my side and grabbed a huge fish head that was used as bait, creating quite a stir in the water as it chomped hungrily and furiously at its meal. It was soon followed by its hungry mates and still nobody in the cage tried to stroke one of them. There was definitely no need for the captain to warn us, I thought and shuddered at the same time but you always get one, don't you? And the one would have to be a nutter or have a death wish. But it was quite an experience, another first.

Believe it or not, the next day on the beach, my wife went in for a swim and it was only later that we saw the 'DANGER, SHARKS' sign. GULP! To end the day, my wife found a mobile phone on the beach and dialled a number to try to locate the owner. It was a London UK number belonging to the boss of the owner of the phone, a grateful guy who turned up at our B&B with a nice bottle of red and who we invited to dinner, happy that his phone had only been dropped and not spat out by the shark that ate him.

NIGERIA
1997

(Fresh meat in Scary Land.)

This was a country that I'd heard various stories about and I must admit, I was not terribly impressed with them. However, needs must, so I was down in London to get a visa at the Nigerian consulate - no problem except for the waiting, which took about two hours, then it was fine details such as needing several injections and the required HIV test certificate, which was mandatory. My feelings were: 'Blimey, no one coming here has a to go through all this!' But what could one do if you needed a visa? A couple of days later, I was on my way to the Liverpool School of Tropical Medicine for an HIV test and various injections. I had a good look around the place before I was jabbed and probed and what shocked me were the photographs of diseases that you could catch whilst visiting West Africa. Some of the images were absolutely awful and kind of put me off having my lunch. However, I got through all this and was free to venture into the darkest place in the Heart of Darkness - Nigeria.

After an overnight stay in Lagos, I took a taxi to the internal airport with our African agent, who soon informed me that my flight was cancelled due to some problem with the plane's engines. I had kind of thought there may be a problem as I watched the mechanics wearing scruffy coveralls. One of the engines had its cowling removed and they had removed a part of one of them. I could just make out what looked like a fuel pump, not the sort of thing you want to malfunction during the flight. So I was not put out too much, as you can imagine, over the cancellation. Our agent, Boy Wonder, informed me that with great effort he could get me another flight to my destination, Port Harcourt, but it would cost me $100. Of course it would, welcome to Nigeria! Fortunately, I had US dollars on me so I bordered the flight.

It was the grottiest aeroplane that I have been on and I was in first class – God knows what cattle class was like, probably the hold. The internal walls of the plane looked hand-painted as there were runs of paint everywhere and the white had turned into a sickly yellowy-grey. I was informed by the agent that one of the local tribal chiefs owned it, so one could imagine that all the servicing and maintenance was up to date and that it had a valid MOT certificate. The engines started after several goes, sounding like a car with a flat battery in January. As they built up sufficient RPMs to move, I admit I have never heard engines sound like these and I have flown a bit in my lifetime. I was rather hoping for some more of that lifetime to be honest but good old Johnnie Walker came to the rescue. I ordered a double, followed by a refill. It kind of quelled the engine noise a few decibels enough to get me to my destination.

I was met by one of the other inspectors for lunch at a local restaurant. At this stage, I was not overly impressed with Port Harcourt but was telling myself I could get through it. My new colleague seemed to know some of the staff. One waitress looked at me and said to him:
"I see you have brought fresh meat with you." I didn't ask.
After lunch, it was on to Shell's office to meet the crowd of only three white inspectors; the others, I assumed, were locals. When we finished the day, it was off to our digs, which was a house outside the town. In fact it was a compound, surrounded by high walls, razor wire and many spotlights, plus two armed

guards on the main gate. Inside, all the windows and doors had iron bars on them and were padlocked, as were all the internal doors and windows. The stories I had heard from the 'when I was in such-and-such a country' type of contractor personnel, the ones who tend to overstate and embellish on past contracts after a beer or two, seemed to be coming true and were not the bullshit I had first assumed...

The first night seemed unnerving, to say the least, as I was told by my two, I shall call 'inmates', to lock and bolt my bedroom door! I had the feeling of 'Cripes, what the f**k is this place?' Or words to that effect. I was to find out all too soon. One night in a local bar, I just managed to get my wheelbarrow through the door (you needed it to carry your Nigerian naira, a currency that's next to useless as you need about 1000 of it for a beer) when a Jock came in, all flustered, and ordered a large whisky. After he gulped it down in five, he blurted out that he'd been having a beer at home, watching TV, when he heard a noise in the back room. He checked but couldn't find any problem and returned to watch the TV. The next moment, two men came through the ceiling of his lounge and pointed their guns at him and they didn't want a beer or to watch the TV! Another 'shaggy' story I was now starting to believe. I got myself an armoured wheelbarrow. No I didn't. I stayed at home and had a beer.

Day-to-day life consisted of a drive to Shell's office to find out if we had permission by the local tribal chief to go to the site which was the oil refinery. This is despite the clout Shell had in the region, but something we had to adhere to. Sometimes it was OK and we were allowed on site (only after a certain amount of dosh was given to old Chiefy) but many days we had to remain in the main office all day, which was quite a pain in the butt. Often, on the frequent drives back to our accommodation, we would be confronted by a lot of zombies, or at least they looked like that: crowds surrounding our car, staring blankly, holding wooden spoons or whatever, for us to buy. We didn't, as we wouldn't get any peace at all in the future but it was not a very nice experience having them staring at us through the windows like some horror film. It took several of these horror films to eventually get used to it - what choice did we have?

One really awful experience happened one day, when a young lad came looking for a job on our site and the next day his mutilated body was found. I was told he had been killed by another tribe. How true this was I'm not sure, but I was not about to investigate the issue. Every day there was something to be afraid of. Sometimes, we had to go by boat to a gathering station. The problem was, they - whoever 'they' were - were taking Shell engineers and holding them for ransom, which kind of made you jittery to say the least. My main concern, apart from the obvious, was that I could miss taking my anti-malaria medication, so I ended up taking a good amount around with me just in case I was one of the unlucky engineers. Sometime I would think, 'What a way to earn a living!' I'd been to some shady places before but this was completely alien to me. A bit different to working in a factory: I don't think anybody would read a book about that. You get on with it. The things we do for oil!

My time was eventually up in Nigeria and I left Port Harcourt without any regrets. On my way to Lagos for my flight home, I first had to fly Tribal Airways again, a short hop but not with a skip and a jump. After I checked in, I was confronted by two baggage handlers who made it quite clear that my baggage would not be put on board the aircraft without a donation to their local charity, i.e. their pockets. Fortunately I still had some naira in my pocket for these problems. I would have binned them anyway as the notes were in such a filthy state. I heard that when the banks got rid of the old banknotes, they burned them and the notes were so soiled with oil and whatever that the furness couldn't cope with them, blew a gasket and burnt down. Imagine burning down a bloody furness! In restaurants, I was only too pleased to pay before I ate but made quite sure that I washed my hands with acid and soap and water before I touched any food. Only kidding about the acid but it did cross my mind.

Back at the airport, I gave both the baggage handlers about $10 each in their currency but followed them out on the tarmac and watched then load my baggage on the plane. I also watched them close the cargo doors like a trustworthy person that I am not. Surprise, surprise, there were many generous, helpful people willing to carry my baggage to a taxi, which seemed a good opportunity to offload some more germ-laden banknotes. On route to Lagos International Airport, my taxi was stopped by the Nigerian plod and a gun was pushed through the taxi window and up my nose and he wasn't, I gathered, going to wish me bon voyage. So I thankfully got rid of the rest of that awful currency. Fortunately I was in business class on a BA flight back to normality and away from Scary Land, never to return!

AZERBAIJAN
1998

(Killer sea, killer dust, killer water, killer mines and a whole new meaning to 'Goat's Cheese'.)

Azerbaijan has got to be one of the strangest places that I've been to on my travels. The capital, Baku, is the largest city on the Caspian Sea and in the Caucasus region. It is also the lowest at sea level. I was picked up from the airport by a Russian-speaking driver (Azerbaijan had been under Russian rule). I could only speak a little Russian, so conversation was rather limited to say the least. It was almost midnight, so I couldn't see much on the way to my next home. We were billeted in quite comfortable bungalows, one to a bungalow which is always good. The project installation was of oil booster pumps which were huge as they had a great distance to pump the oil from Baku to the Black Sea coast stations for export.

On my first day, I took a look round the site which wasn't anywhere near completion, mainly due to a lack of spare parts or any parts at all. As one can imagine there wasn't a hell of a lot to do. I thought, 'Shit, this is going to be a bore,' but got on with it best I could. The first thing was to inspect the site status and try to find out what was holding up deliveries, then meet the local engineers and technicians. The team from the UK consisted of four engineers: mechanical, instrument, electrical and operations. I was the mechanical one.

On each daily trip to the plant, we had to traverse a railway crossing, manned by an old woman and probably two of her grandchildren who looked about eight years old, so we were rather glad the trains were not frequent. On arrival at the plant, I was given the keys to a 4x4 Lada, which had a mind of its own. Put it in first gear it would reverse and the clutch did nothing but helped to wear out the gears, if they could be found. A larder would have been better to drive. Basically it was a death trap, so nobody wanted to drive it, especially yours truly. So, soldier on we did, minus the Lada. After dinner I was invited to go into Baku for a few drinks, which happened to be Turkish beer – probably not the best lager in the world and it tasted full of formaldehyde, definitely not Carlsberg but did the job.

As you drove into town, you noticed oil pipes and valves running alongside the houses on both sides. There were nodding donkeys (oil pumps that you see on TV in oil fields in Dallas and places like that), the difference being that in the US the pumps were around the oil fields and not the houses themselves. Also shocking were the large oil spillages dotted around the place but, and this is a big but, there were some fine historical buildings, almost like castles, giving Baku a look of mystique. It seems rather strange to hold a Formula One Grand Prix there. Good old Bernie Ecclestone would no doubt have held one on the moon if he could!

The site ran alongside the Caspian Sea. I drove down there one day to take a look (not in the Lada no-wheel drive) and couldn't quite believe what I was confronted with. The sea was a strange green colour, somewhat radioactive, and beached there were several scuppered ex-Russian naval vessels. That weekend, I returned with my swimming costume, not for a dip but to get some sun and to get away from our base. There were many locals swimming there in Russian Speedos (although I doubt they're still alive). I got talking to some

German environmentalists who were analysing the Caspian Sea, who warned me not to go near as the readings went off the scales on their equipment and there was a disused chemical plant further down the beach pumping waste chemicals in it. Yikes. No wonder some of the locals looked a bit green when they came out. The German environmentalists and I had lunch together. I had some bread and cheese and they informed me afterwards that the cheese I'd just eaten had been cured in the stomach of a goat, *while the goat was still alive.* Why did they not mention this before I ordered it?! Fortunately, I was OK and it tasted fine. I actually recommend it. Just don't go for a swim afterwards.

Across the road from our site were some huge buildings. The electrical engineer and myself ventured inside to take a look. They were obviously abandoned iron foundries; one building must have been a mile long and was rusting away. It was full of old worn-out machinery and all kinds of scrap: rusting deteriorating iron pipes, gear wheels, large iron plates. It was obvious that all the good and useful stuff had been pillaged by the Russians who had upped and away. That's their business, hey? As we walked around, gobsmacked, we came across an outside cafe that sold beer. We had discovered a watering hole, a bit strange but what the hell? It was quite pleasant sitting outside under the trees with the sun shining, having a few beers before dinner. A pleasing ambiance, considering where we were.

We used to drive to another site, some 20 or 30 miles from Baku along 'Death Road'. This road was a nightmare. Old trucks and heavy vehicles would pass on the other side and miss you by inches. Strangely, these vehicles didn't carry spare wheels but spare inner tubes hanging on the arm of the rear mirrors. I was told that the tyres had very little tread on them, mostly none at all, so when they got a puncture they'd replace the tube only! Scary or what? A year or so after I left this project, I was told that the electrical engineer was severely injured in a head-on collision and very lucky to survive. One of my nine lives saved by proxy!

One day while driving down Death Road, we stopped for a coffee break in a small village. I couldn't help notice that the woman running the cafe was washing the dishes in the gutter as it had been raining. Even more off-putting was the goat urinating alongside her, straight into the gutter. That ended the thought of a coffee break. We carried on... Next to shock us was the drive past a disused smelter plant. Red dust had settled on everything the eye could see: the surrounding houses, the trees, cars, animals. It was heaped in mounds all over the area. The dust was bauxite mixed with caustic solutions. [This red mud is further processed to produce aluminium oxides; the waste product from the process is called brown mud but is red in colour. Bauxite refineries produce alumina ($A12\ 03$) which is primarily a feed stock for the aluminium-reduction industry. Waste muds are created by the extraction of alumina from its ore.] Whilst bauxite may contain levels of enhanced naturally occurring radioactive materials (TENORM = technology, materials, enhanced, naturally occurring materials), it may contain primordial radionuclides or radioactive elements. Nice stuff to leave around the place! No wonder that we held our breath and closed the windows as we passed here! No wonder the world is in such a state!

We arrived at the base and were shown to our rooms which, considering the state of the surrounding area, were great. The food in the restaurant was excellent too, basic Russian but good. As we didn't have much work to do at our own plant due to lack of parts, we had been tasked with surveying the area prior to the Azerbaijani engineers erecting electrical pylons. When in these odd places, one can get involved in bizarre work that's out of one's scope - but better than being

stuck at our plant, bored stiff! After breakfast, we set off up in the hills looking down on Georgia (home to Eduard Shevardnadze, the late Soviet politician and diplomat). I was climbing up the hill to see the view when our driver shouted:
"Hey, mister, mister! Be careful!"
"Of what?" I could see old and broken Russian tank parts scattered all over the place.
"Mines, mister! Mines!"
"What. The fuck." I carefully retraced my steps as best I could before continuing. "You're saying this is minefield that hasn't been cleared and you didn't think to tell me this before?" The question went over his head. There would be some explaining to do back at the base! Somehow I felt more relieved than angry. What is it about me and mines?! Anyway, I was still in one piece...

So after surviving a minefield, we headed back to base in Baku. It was our last night and we were treated to a slap-up dinner in one of Baku's finest restaurants, resembling an old castle or fort. I could only assume that they felt a wee bit guilty about almost sending us to meet our maker. To sum up Baku and Azerbaijan, it seemed as if I'd arrived there after a hundred year war had ended. But today they host a round of the Formula One season there. Strange world, hey? Better than being bored... JUST!

Egypt 2nd Trip 1998

(A bomb, some mines, one dangerous pair of trousers and no place to pray.)

The last time I was in Egypt was to obtain a visa to get into Saudi Arabia more than twenty years ago and Cairo was hot and smelly. It had remained the same today. I stayed overnight in a touristy hotel – this was when terrorists were blowing up ancient monuments in the Valley of the Kings. I was working for the French oil company Total which should be interesting as I spoke very little French - well, enough to order a beer, which turned up as a whisky, and my well-done steak turned up the colour of a Manchester United FC shirt, still running around the restaurant with the waiter chasing it. The site was a gas plant not far from Hurghada, the scuba-diving resort on the Red Sea coast. We set out early the next day from Cairo, three hours in a car minus air conditioning. On arrival, I was met by the French superintendent who introduced me to my motley crew of Egyptians; most could not speak any English. This should go well! I was going to have get my French and Arabic going otherwise I'm dead before I start. My Arabic was passable but my French - umm, well, I could now order a raw cow. My room was… let's say 'passable' but I would only be on duty for a month at a time as I had a 'back-to-back'. He turned out to be a Welshman and I don't speak that lingo either. This was going to be fun!

I spent the first day looking around, finding my feet. I could tell the place hadn't received much maintenance judging by the state of the machinery but my job was to get it sorted. To get it sorted, I needed a good crew of technicians and engineers, so I headed for the main workshop to speak to them. I didn't get much joy as they all seemed half asleep. I made a start on who's who and what's everyone's position? As soon as I was told all their names, I immediately forgot them as one does. There were three supervisors, one electrical, one instruments and the other mechanical, so I had them show me around the plant to ascertain how knowledgable they were and how they'd been maintaining it. After the first few questions, I gathered that they knew sweet Fanny Adams about the place and even what it was for. How the French oil company hadn't realised the situation they'd put themselves in I do not know! However, it was too late to turn back now. So here I was, working on a potential bomb with a motley and fairly bewildered crew. 'Beam me up, Scotty!'

Lunch was also 'passable', after which I went down to the Red Sea, which was just outside the perimeter fence, and spent the remainder of my lunch break thinking and planning.
"Hey, mister, it is too dangerous to walk on beach!"
"Why?"
"Because army have put mines down to keep the Israelis out!"
"Oh how nice to know.'
So that put an abrupt end to my lunchtime plans. I didn't see any warning signs displayed anywhere. 'How strange,' I thought. Still, by this time in my career, I'd survived being blown up in more ways than I'd like to think!

Back to camp was an hour's drive in a small bus. I showered and got ready for dinner. My trousers were creased to hell but I'd spotted the camp laundry not too far from my room so went to get them pressed as I wanted to make a good impression with my new French colleagues. After waking the Egyptian presser from his tiring day of lying down asleep, he got up, grabbed my trousers, laid them down on the ironing table, took a swig from a plastic water bottle and spat it out, covering every inch of them and proceeded to press. Mental note: do not put hands on trousers during dinner. Or ever. It was going to be difficult putting them on.

Dinner started with a few beers which was most welcome, followed by saucisson (a French delicacy literally meaning 'large sausage') and a decent meal. At the end of the first day, I had a quick check on the equipment/machinery and all seemed to be working OK. My next task was to get a maintenance schedule up and running and organise the crew. The planned schedule was no problem but getting the crew to work was. After lunch one day, I walked into the workshop to find them all sitting down minus their shoes and socks. I had to put a stop to this right away and make it clear that I was a boss that didn't stand any nonsense or I'd lose control. Very often it was a case of gaining respect or a knife in the back. "Move your flipping bottoms and get some shuggle (work) done!" I shouted in my best Scouse.

It turns out they were intending to wash their feet and pray. This was an adjustment for me being an atheist, quickly learning how things are in many parts of the Middle East. Religion plays a major role in people's lives and who am I to question it? After showing them basic engineering and maintenance (as praying is no good on its own), I had an idea. I'd found that the chief electrician couldn't change a light bulb but was a brilliant carpenter so I called him Joseph, which won him over, then I sent the entire crew on a mission to gather as much wood and build their own mosque. I figured that if they wanted to pray, they might as well have a proper place. I don't mean a huge place, just a wooden surround and a roof, but they were on cloud nine and even brought some prayer mats and holy pictures. The work rate improved immensely, although I had to keep an eye on what was still essentially a bomb staffed by people who barely knew how to run it. And me with no god to pray to.

The French were clearly thinking: 'Who is this crazy roast beef?' and, no doubt to test my sanity, the French plant manager asked me to organise and supervise fire drills. So I got the crew to prepare the equipment outside the perimeter fence, away from any nasty stuff on the other side of the plant that they'd never be able to extinguish and, of course, the possible minefield. Then I stood well back and watched. They laid down bits of newspaper and a few twigs and lit the fire, then each one had a squirt on a fire extinguisher. Of course their fire was out before they even got to it. I decided to cancel the drill, and get the crew to fetch bigger pieces of wood and some old tyres from the local tip and some petrol. The next day, we set our fire as near to the gas plant as we dared, whilst still being far enough away from the mines, poured petrol on the lot and lit it. Whoosh! By the look on their faces, they had obviously never been to a Guy Fawkes party. Well, no point having a fire drill with a flame the size of a candle! Unfortunately, the other Total plant on the other side of the Red Sea thought it was a real fire and began phoning our plant operators. OK, I was a bit ambitious but at least it raised some action and the crew learned what a real emergency could be like!

Fire drills and mosque building out of the way and it was down to business. Things started to turn out the better side of bad and the crew, apart from feet washing and prayers after lunch, succumbed to work. I quite enjoyed my time there. Then a long drive to Cairo and a relaxed BA flight home. It wasn't the best of contracts as it only lasted two trips. Contracts can't all be good but it keeps the wolves away from the door...

Norway
1998

(In which I avoid death in a glider and 'exchange pints' with a local.)

Not the most exciting country I've been to but it is very beautiful with such gorgeous scenery. I flew in from Aberdeen Airport on a very cold and stormy day, what else?! The aeroplane was a small charter so it took a bit of a battering going across the North Sea. The landing did not do my heart any favours as one could almost touch the rocks just in front of the runway but it made it, just. From Stavanger, I got the ferry to Haugesund. The company I was contracted to was Kellogg Construction UK, who were commissioning the gas plant for Statoil, Norway. I was booked into a hotel until I could find permanent accommodation. The hotel was cosy and, as it was December, the restaurant had all these red and green plants all over the place, all Christmasy.

The project was on the largest processing plant of its kind in Europe, which was an hour's drive from where I was staying. Statoil had a coach laid on to ferry us to and from the plant. It soon started to snow, which made the place much more beautiful. What amazed me was, most of the houses were made of wood and nearly all had lots of candles lit up inside them every evening. Though I did come across a house fire, nobody seemed too bothered - except, I would imagine, the owners that were left with a heck of a lot of ash! And of course the bill.

It didn't take long to find accommodation as there were lots of homes with spare rooms available for rent. I chose one with a basement consisting of a lounge, a bedroom, a bathroom and a kitchen, owned by a local GP and his wife. You could see the ships on the North Sea from the lounge window. It made a refreshing walk on a Sunday morning (or afternoon) after a Saturday-night visit to their local bars. I would walk into town, about a mile away, and be greeted by lots of parties going on. Alcohol was so expensive in Norway, our Viking friends would be drinking their home brew and only venture into their local bars much later on in the evening to avoid the large bills (and add the spills, as I was soon to find out).

One particular evening, I was standing in a very empty and quiet bar, when a crowd walked (or rather staggered) in and one guy bumped into me, knocking half my pint all over me without an apology, and just walked on and sat at a table with his mates. I decided to pay a visit to the toilet and, blow me and damn, had to pass his table to get there. Unfortunately, I slipped and spilled the remains of my pint all over him but I did apologise, sort of! You may think that I am as bad as him but the lack of an apology annoyed me most - oh and my wet clothes. I ordered another drink and watched my back on the way out!

I was enjoying time in Norway as the weeks were flying by. Some weekends, I would take the ferry down the fjords to places like Oslo and the chairlift up the side of one of their mountains - good fun! Monday mornings, I would start with a walk to the main road to catch the coach to the plant. As I lived the furthest away, I was picked up first and managed to get a good seat at the rear, away from the soon-to-board throng. There's always one irritating type, a 'one company all his life' staff man who likes to be in control and sits up front next to the driver as if he's riding shotgun for Wells Fargo. There were only a handful of agency personnel; the others were all staff. Kellogg had this thing about sending out all their young graduate engineers, no experience of life on site (and life in general) but loads of experience in the office, drinking coffee and waffling! We were basically there to get the job done and mentor their puppies. One of the annoying parts of the job was letting them take the reins at the meeting, with all the worldly/experienced contractors and sub-contractors looking on, and not saying a word. Of course the contractors who knew the game made mincemeat out of them but I was getting well paid, so let them get on with it. I suppose we all have to learn. Anyway, I didn't lose any sleep over it and the project was completed.

The thing I liked about Norway was the fact that it was so clean and welcoming. You felt healthy just being there. I dared not mention Chernobyl to them as they must have received their share of the fallout being close to Russia but nothing was said on the subject. At weekends, I would have a sleep-in and a walk along the coast which was just outside, to clear my head, then a full English breakfast which was a nice treat, then another walk, sometimes into town, then a pint or two at one of the local bars, Sunday lunch, then back to my digs to work on this book. I joined a flying club, which turned out to be a gliding club. I told the instructor that I'd had flyings lessons in a Cessna 150 plane but none in a glider. The glider had an engine to get off the ground and we were soon circling the town, then headed to the coast, passing the main airport. The pilot said, "OK, you take over," expecting me to take to it no bother. I had the controls and tried reminding him about the 'only a few flying lessons and none in a glider thing' but all I got from him in his pidgin English was "no problem!" I looked out the side window and racing down the runway below, about to join us, was Scandinavian DC-3. It didn't seem to bother him but I thought I'd better point us away from the airport. "Why are you going starboard when our runway is the opposite?" he asked. I pointed out the whopping great bird about to hit us up the chuff! "OK, I take over for the landing." Another life gone but a great experience.

One day, when this damn bloody Coronavirus goes away, I'd like to return to Norway. I'm writing this in July 2020 and it looks like it's going to be with us a bit longer. I am looking more like Ben Gunn from Treasure Island as the weeks pass... A few thoughts regarding Covid-19/Coronavirus: Trump is a bloody idiot. How can one say that this came from China and we all should drink bleach? I was introduced to some of Trump's supporters once at breakfast in a hotel in Italy. I couldn't help myself and said: "Welcome to World War Three!" This pissed them off rather. Fortunately it didn't happen but how close can we get?! All I can say now is 'Thank fuck he's gone,' as we stick to the fat man with the unruly mop telling us when we can roam about again. I do feel that we are not alone in this universe, otherwise intelligent beings would have found us by now! So sod off, Covid, and give us a break.

USA
(& Monaco and Brazil)
1999

(Where I nearly get eaten whilst drinking, eyeball a Prince and take an important cab journey.)

America. Well, what can one say about this place?! But here goes...

It's big. I first visited it fifty-odd years ago when I was a seagoing junior marine engineer officer, starting on life's wonders. Going through immigration was a nerve-wracking experience as I was subjected to a barrage of questions such as: 'Are you a communist?', 'Have you ever had communist tendencies?' and 'Ever made any bombs?' All this for a twenty-year-old rookie first time overseas. On a recent trip to New York, prior to boarding the flight at Schiphol Airport, Amsterdam, I was subjected to this again. But alas, you can't blame the Americans for subjecting visitors to these questions after 9/11 and with the world in such an unpredictable state, can you?

An FPSO (as you may have read in other chapters) is a vessel that has either been converted from a supertanker or is a new purpose-built vessel. It is bigger than a supertanker due to the fact that it has a production plant on deck with all sorts of machinery, pumps, gas turbines, generators and many large tanks and vessels. FPSO stands for floating, production, storage and offload. Basically an oil rig sucks up oil from its oil well then transfers it to the FPSO. The FPSO then separates the water and gasses and produces oil, which is then stored and pumped/offloaded to a supertanker that is now on station. The gas is then pumped onto shore and the loaded tanker sails off to the buyer, which is mainly America!

I flew into Louis Armstrong Airport in New Orleans and was greeted by John, my new back-to-back, then it was a short drive to the nearest car-hire place to pick up a car that I was to use for the duration of my time on this project. I couldn't resist hiring an American hot car, my first time in an automatic! Then it was a drive to our base, a place called Houma, Louisiana to check into our hotel, a Marriott. John took me to a bar across the road so we could get to know each other and I could learn about the project, and the next day we drove to the shipyard which was about an hour from the hotel. My new American colleagues seemed a friendly bunch, easy to get on with and very helpful, mainly Cajuns from the bayou areas of Louisiana, descendants of French Canadians who spoke an archaic form of French but it all sounded American to me! One in particular rather fancied himself as John Wayne. I soon nicknamed him 'The Dook' (Duke, as in 'of Hazard') which he quite liked. His vehicle was a big pickup truck with huge wheels like a racing car with big tyres (or tires in American speak). I wouldn't have fancied the petrol/gas bill but hey, that's why there are huge supertankers coming and going across the oceans to fill those unnecessary guzzling vehicles! When will they learn?!

The job was to supervise and inspect the equipment being installed and complete a final inspection. A huge road crane arrived on site, the ones that you see on building sites but bigger (we were in the USA, after all, where things tend to be bigger). This was to be loaded on our vessel to be used on site in Africa to move equipment about before the ship's deck cranes could be powered up and load tested. It was one of my jobs to witness the tests and sign for it but of course the crane engineer didn't have a driving licence for the USA, so yours truly had the job of operating and driving this monster. The easy part was to operate the crane: push lever back: crane jib goes up; push lever forward: jib goes up and so on. I was getting used to this, easy peasy, but one gets overconfident, doesn't one! So when it came to road-test time, I said, "No problem," and off we went.

I waved to the big security guard on the shipyard gate as I always did and a huge smile was reciprocated. On the return, I gave a wave and a smile to him but this time his smile was a grimace and he was pointing to the main gate that used to be there. 'Oh shit, what have I done?' Of course I'd knocked the bloody gate down is what I'd done! And hadn't felt a thing in this monster crane. Next thing I knew, I was taken to the shipyard clinic for drug and alcohol tests. Fortunately, I hadn't drunk the night before and don't take illegal drugs so that was that, just a few minor scratches on the side of the crane and a red face for me. Of course it took a while to get over the ribbing that I got too! They were not pleased at our main office in Monaco either; somebody had to drive. So the moral of the story is: be careful what you volunteer for, or just don't volunteer.

One weekend, three of us hired an air/fan boat to sightsee on the Everglades. It was a great trip with lots to see. These flat-bottomed boats skimmed the water and a lot of wildlife lived in the shallows. The boat was powered by a diesel engine about the size of a truck engine, coupled to a massive fan which propelled it along the water at about 30mph. We skimmed for about an hour then came to a patch of land. As we approached, I heard this loud splash like a tree had fallen in the water. The driver assured us that it was no tree but an alligator! We tied up the boat and I followed him on land. He soon pointed out an alligator's nest and proceeded to pick out some eggs whilst informing me, as I stood next to the nest, wishing that my colleagues would hurry up and take a bloody snap before the alligators did, that the loud splash we'd heard was likely to have been Mummy or Daddy out to protect their brood in the nest and was probably watching us, waiting to pounce. As this information was being absorbed by the gearwheels and cogs in my head, the cunning bastard driver grabbed hold of my ankle and with the newly acquired and absorbed information about parent gators - well, one can imagine how quick my reflex actions were in getting back to the boat. *(* see the actual snap of the snap below.)* 'Nuff said! But a fun day.

I was beginning to enjoy working in America as it was both so different and yet familiar after seeing lots of American TV back home. On some nights, we headed to a karaoke bar where I gave my rendition of Frank Sinatra's 'New York'. Well, it gave patrons a laugh; I must have been a bit pissed to say the least. Another bar I frequented was Hoots and Coots, a bit of a wild place but this was Louisiana! 'Hoots' was an actual person, a larger-than-life character - literally in fact, as he was rather fat but a good soul who was knocking about with a female tug-boat captain, quite a lot smaller than he was. A tug boat and a life raft. There were some rather odd people who drank there and some local lads in their twenties but Hoots used to let them in; he didn't give a hoot. One night, I was standing at the bar having a beer when something caught my eye. It was an alligator about two foot long that had its eye on my ear, not my beer, and was heading my way along the bar top. This was the local nutter's party trick to unsuspecting visitors. The sods had sneaked out the back door and bought back this baby alligator from the bayou that was by Hoots' back garden. It doesn't half sober you up when one of those walks by on the bar top. I didn't stroke it but thought it best to get a table just in case it was hungry.

My home leave was now due, so I combined it with yet another offshore survival course in Morecambe, UK, which was also due as you cannot go offshore without a valid survival certificate. I used to hate having business duties on home leave but what can one do? Home leave and business completed, I flew back to New Orleans, then caught a taxi to pick up a brand-new rental 4x4 which made a pleasant change. It was now dark and not the best time to drive in a strange country. Also, Hurricane Katrina had been and gone, the roads were still flooded and conditions were not too good to say the least! On some roads and freeways it was like driving in a lake - Katrina had left lots of debris scattered all over the place and caused quite a lot of damage, even from what I could see during the night-time.

It was getting near the end of our project in Houma and personnel were leaving. The night before they departed usually ended up in a visit to the bar across the road from our hotel to wish them a bon voyage. This particular night, I didn't feel like mixing with the drunken horde, so I set off on my own for a quiet drink. On my return, I was just about to enter the hotel car park when I stupidly thought: 'Sod it, I'll have a nightcap and say goodbye to whoever's leaving.' Not a great idea but at least I could leave the vehicle in the bar car park if I had more than a nightcap. After two beers I decided to leave as there's nothing worse than being in the company of drunks when one isn't! It was getting a bit out of hand as they'd been at it all night and not good coherent company. The cheerleader for the night was our project manager, a short and nasty-minded Jock who I could never get on with. He spotted me leaving and asked me to take him and some of his Jock mates to the local lap-dancing club. Foolishly, I said OK. After an hour in there, I decided to leave and they all piled out and into the 4x4.

Houma is a small town and 3am is not the best time to be on the roads with a bunch of drunks in the back. I wasn't drunk but would not have passed a breathalyser test which would have ended in an overnight or several nights in a downtown Louisiana jail (I have seen the film 'Cool Hand Luke') plus a huge fine and no chance of any promotion in the company or ex-company! I drove along as steadily and as quietly as I possibly could until the little Jock put his feet on the dashboard and started kicking at the windscreen! Fortunately it didn't break but the electric rear-view mirror was hanging by the wires. He then started opening and slamming the passenger door very hard, shouting and making one hell of a racket. Had the law been around, we'd all have surely been shot dead or worse!

We arrived back at the hotel, where I parked up then saw the damage and remembered this was a brand-new top-of-the-range 4x4. I shouted at the little shit to get out, whereupon he stuck his chin out and said, "Punch me right there." I really don't know how I kept my hands off him as I felt that I didn't need a promotion but one of the good Jocks, who I did get on with, had a firm grip of my arm. So he lived to piss people off another day. I told him, "You had better get it fixed, you little Scottish c**t," and went to my room, pissed off with myself for allowing this to happen. The next day he came into my office and made a feeble attempt at an apology. I told him where to go and he stormed off cursing me.

On Saturday, I drove to New Orleans to collect my wife from the airport for a short stay. I had booked a table at my favourite restaurant in the town but to my horror, the Jock had booked several tables for all the crowd so we went into the next room. He spotted us and sent a waiter to ask us to join them. I refused. He even tried to pay our bill but again I refused the offer. Was I becoming a pain in the arse and ungrateful? Fortunately, my wife had a good time as she'd been shown around by some of the local friends that I had made and taken around their shrimp boats and factory.

Normally I would forgive and forget but he was such a little shit that I couldn't.

I had mixed feelings on leaving Houma. I had enjoyed my time there until Jock Strap appeared again on the scene. I was preparing to leave for Angola to join up with the vessel, only to find that the Scottish git was on board! I thought: 'Somebody's taking the piss here!' But no, he was running the project. It was not long before we clashed yet again. I was witnessing the slewing and weight testing of the ship's main cranes for certification and insurance. Before we commenced, I radioed the control room for them to give out a warning announcement as to what we were about to do, as we are talking about tons of sea water in huge canvas bags that were about to be dangled over personnel's heads. Of course Jock didn't like this as it would hold people up for a few minutes and of course I took no notice of his radio messages. Next thing I knew I was on board the next chopper out to the mainland and instructed to fly back to our main office in Monaco. I had no complaints.

We touched down on a remote airstrip to wait for the plane to take us to Angola and the international airport to catch the the Air France red-eye to Charles de Gaulle, Paris. After a long wait in a hot and sweaty hanger with a large crowd of what looked to me like an army of war-torn mercenaries (there were lots of them there during Angola's wars), an old Russian army plane landed and the throng started to run towards it with its engines still running. Me too. I managed to clamber on board and find a seat with a huge map table in front of me. This plane had obviously been used by the mercenaries during the troubles but that didn't bother me so long as it took off and stayed up.

At the airport in Angola, I was rather pleased to have a business ticket on Air France which gave me much-needed access to the business lounge. There was to be another hitch, however. Instead of Air France, I had to board a TAP Air Portugal flight to Lisbon. 'Ah well, anywhere but here,' I thought, ordered a large Scotch and proceeded to read an old newspaper and catch up with the world. We'd been going for about an hour when I heard somebody shouting and a woman crying. This time I thought, 'Oh shit, what the…?!' I looked out of the window and could see sparks coming out of one of the engines. There was nothing I could do but sit down to finish my drink and quickly order another just in case circumstances dictated that the bar was to close. The captain came on the intercom to say that we were experiencing a problem and had to return to Angola. He avoided telling us to brace, the universal indication we were about to die, but I had a gulp of my Scotch and thought: 'Well, it's his job, he can do the worrying,' and ordered another.

We eventually got back in one piece and were stuck in the business lounge all night. I managed to grab some sleep on the carpet (good job it was the business lounge and the carpets were reasonably clean) and was bussed the next day to a hotel. The Air France flight was 1500 hrs. This was still 0600 hrs but I dared not get into a deep sleep so dozed for a bit, had a shower to stay alert and went to get some breakfast, lunch or whatever the hour dictated. Then back to the airport to catch the rescheduled Air France flight. I don't normally sleep on planes if it's not a turn-left-at-the-top-of-the-stairs seat with a bed to sleep in after one or six champers but I sure did this time with a 'please do not disturb' sticker on my forehead.

I boarded a flight from Paris to Nice, then getting off the Nice-to-Monaco chopper, I walked to the Hotel de Paris, Monte Carlo, as it was quite near the heliport and another of my watering holes when I was based in Monaco. I went into the bar. The usual barman was there and we were discussing my latest escapade over a few beers and the many different types of vodka on display when I noticed a bottle with a clock. He told me that it was a rare Napoleon brandy that was due to go on auction and gave me a slug of it when nobody was looking. I found out at a later date that it went for $5,000. I mean, it was very good but...?!

Back in our Monaco office, I got an update on events. They were sending me to the Rolls-Royce factory in Sao Paulo, Brazil, to witness their engineers dismantle and strip-down a gas turbine that had thrown a blade. Basically, the turbine had been doing over 3,000 rpms, when a turbine blade had worked itself loose or fractured and gone through the engine causing - I would well imagine a considerable amount of damage. It had already put the FPSO out of action as the turbine was the main generator. Fortunately they had a spare turbine on board to install but then had no spare in case of any further problems. So the next day, I choppered back into Nice to catch a flight to Brazil via Paris and settled down in business class with a glass of wine to update myself with some information on Rolls-Royce gas turbines. There was plenty to cram in. R&R had done good and sent a driver to pick me up at the airport to take me to their factory in the centre of town. During the drive, I imagined the late great Brazilian racing driver Ayrton Senna passing along this road on his way to the airport to fly to the Grand Prix...

At the factory, I was introduced to the manager and his engineers and was pleased to find out the place was closed on Sunday which gave me a day to get un-jet-lagged and do a spot of sightseeing. The two Rolls-Royce specialist engineers were arriving from the UK on Monday to carry out the strip-down. I didn't venture too far as the hotel was in a rather seedy area of Sao Paulo. On Monday morning, I attended a meeting with the Rolls-Royce brass and met the two engineers with whom I was to work. The engineers and I had a proper meeting that night in a local restaurant over dinner and a few bottles of wine. We soon got on well and formulated a plan of how they were going to tackle the job ahead. We had separate offices to the R&R guys, so when they went off to start proceedings I quickly followed them, so as not to miss anything. On the third day I didn't catch them going to the workshop but eventually tracked them down, looking at what was left of the defective turbine blade with grim faces. To their annoyance, I kept hold of the blade, which was coming back with me to Monaco!

The next step was to catch a flight to Mount Vernon, Ohio, to visit the Rolls-Royce USA factory and inspect the power turbine there. There wasn't a driver to meet me at the airport this time so I had to hire a car. I got some directions from the car-hire firm and drove on. It was now dark and the rain was pelting it down. I was driving an American car along a freeway with very big trucks driving up my chuff; they were literally inches away and sounding their huge horns. It would have been crazy to carry on so I pulled off the freeway at the next available exit and soon came to a strange motel which looked like every motel in an American horror film but who cared? I was away from those looney truckers and out of what was now a storm for the night.

Just my luck, the restaurant wasn't open, so I made do with a vending-machine coffee and an early night. It wasn't quite as easy as that because I had to get change from the receptionist who was Mexican whose English was as good as my Spanish. The next morning, after one more vending-machine coffee, I set off to the Mount Vernon factory. The storm had subsided and the day was fine. I soon found a cafe and had breakfast and a real coffee which set me up for the drive, during which I couldn't help noticing that almost every house had the American flag flying outside on a pole. At the R&R workshop, I started to inspect the power turbine but couldn't find fault with it so it was on to inspect the turbine engine in Sao Paulo.

This done, I returned to Monaco to report my findings which were eagerly awaited. The fault was in the casting of the turbine blade which I could see with the naked eye and, looking through some old reports, this had occurred before with parts from the casting company from the same batch. However, this had to be proven because it was now an issue between Rolls-Royce and ourselves. They were saying that it was our fault for our own crew misoperating the turbine on the FPSO. I needed some definite answers fast as they were sending over their lawyers to Monaco. I knew I nearly had the problem solved but proving it was not going to be a walk in the park.

I flew back to the UK and visited the Rolls-Royce main factory in Derby for discussions but, to no great surprise, nobody wanted to admit that it was their fault. The next step was to pay a visit to the Welding Institute (TWI) in Cambridge to use their electron microscopes to inspect the faulty blade. We put the samples in the electron microscope which I'd had experience with at the University of Damascus metallurgy laboratories in Syria, and sure enough I found a definite fault in the blade. It was porous and you could see the interstices/small spaces in the material that would surely allow liquid or gas to pass through. 'BINGO!' No wonder it had failed rotating at over 3,000 RPMs! So it was back to Monaco for debriefing before facing the Rolls-Royce lawyers who were arriving the next day.

That night, I went for a walk up the hill to Monte Carlo to gather my thoughts before the next day's confrontation with the silks. Just before I reached the top, I sat down for a break and a think when a cavalcade of big black cars appeared. The one in the middle stopped a few metres from where I was sitting and a familiar face stared at me through the window. I immediately recognised Prince Rainier III who had ruled Monaco for 56 years, the longest European ruler in history!

For a brief moment, there we were, just staring at each other. I wondered what he was thinking and wondered if he was wondering what I was thinking. I somehow doubted that as he drove on. I reckoned that he must have had more important things to do than to take me for a pint in his local! Perhaps another time, Mr Rainier, but it would have been fun.

The silks arrived in force the next day and we gathered in our conference room. I could sense by their stuffiness that this wasn't going to be easy and it wasn't. There was a lot of money and a lot of reputations on the line, including mine! I spoke about my findings and lab results and it was left to our senior management to come to an arrangement in order to get on with future business. Oh, we won the case! So it was back to work again.

My next trip to America was to work on a repaired oil rig that was in the middle of being towed to Angola from a shipyard in Texas when it had the misfortune of colliding with another vessel and had consequently half-sunk, which didn't do it the world of good - or indeed the world the world of good. At the time, I was on another project and contracted to Sonangol, the oil company of Angola. I was tasked to fly over to the US and inspect the damage, no easy job but glad to get out of Africa. I arrived in Houston late afternoon and was greeted by Taff, another Welsh ex-colleague who I'd worked with before on a project in Gabon. It was now too late to get to the office so - what else? - you go for a beer and chat about old times, oh and the sinking of the oil rig and its current status to bring me up to speed. Even though I was jet-lagged to hell, we visited several bars - one in particular was like being an extra in the 80s TV show 'Dallas'. It was full of JR Ewings in their big ten-gallon cowboy hats!

After a week in the Houston office, I flew down to a place called Aransas Pass, where the oil rig had been towed into a shipyard for inspection. At the yard, I was introduced to their project manager. As I went to shake his hand, I realised there was no hand to shake but a withered right arm. Too late, my hand was already out and I shook his stump as if I'd done it a million times before. Aransas Pass seemed just as barmy as Houma. The office and site personnel were die-hard Texans: even their safety hard hats they wore on site were shaped like ten-gallon hats. I, for one, was not enticed to don one, even as a bet! I was staying in a three-star motel, comfortable but definitely not the Ritz. One night I decided to go to a local bar as you do when you are alone in a strange place. "Be careful," the barman told me. I'd left my car at the motel, I explained, but he was telling me I could be arrested by the police for walking after having consumed even as few as two beers. That's how strange America can be to us Europeans. Another strange thing was, when you visited the site, you had to wear your safety helmet even inside your car. Imagine leaving an air-conditioned office and climbing into a car that had been outside in the Texan sun for hours and having to put on a safety helmet! But that was their rules. Even though one's head was jammed against the roof of the car as you drove along and you couldn't open a window because of all the dust. Still, when in Rome...

After a few beers that night I was soon in cloud cuckoo land, only to be woken by something crawling on my face. I swiped it off and put on the bedside light to find a huge cockroach making a desperate attempt to survive my shoe! So next day, it was a swift exchange of accommodation. I found a smashing hotel right on the waterfront where a certain chap by the name of Michael Collins had stayed prior to a certain historical Apollo 11 moon landing! My wife joined me for a vacation and we were soon riding horses on Mustang beach. That was better than cockroaches trying to eat one's tonsils. At breakfast, the manager of our hotel used to whisper to us as he pointed out, in his words: "The Four Waffle Women." These were certain rather large breakfasters who could eat four waffles for starters. The local supermarkets were full of them, all driving 'Fat Mobiles,' scooping up their fatty food from the shelves into baskets without moving their fatty backsides. 'That's America!' I thought, until I got home and saw the British versions!

New York: On my second trip to the Big Apple in 2012, it was to see my eldest daughter, Elizabeth, an actress, onstage in Brooklyn on the final leg of her world tour for the famous Cheek by Jowl theatre company. We were joined by our youngest daughter, Sarah, and her boyfriend, Patrick. After the final performance, we got into two NYC yellow cabs. I got into the front one with Patrick and, as we drove along, he asked me if he could marry Sarah! I had no hesitation in saying yes as Patrick, who had obviously planned all this, is such a great guy. We arrived at Grand Central station, straight to the bar, ordered champagne and left it to Patrick to finalise his plan, which fortunately went well and now we have two fantastic grandsons, Leo and George! New York is some place; we did all the usual sightseeing and even visited Patrick's relatives in Brooklyn.

That put my American chapter on hold until some years later when my wife and I went to Montana for a holiday stay on a ranch, which was incredible. We did a lot of horse riding on trails and corralling cattle into their pens, plus the food was marvellous. No wonder some of the cowboys worked for free! Of course, I almost came off onto my arse. We were passing some trees when I brushed against a dead branch which made a loud crack and the horse bolted. I fell forward, grabbing its neck and holding on for my life, with its ears going up my nose. I managed to stop it before I died! Back at the ranch it was a shower, a cold beer by a log fire and a super dinner. We next took in Yellowstone park and stayed overnight atop an active volcano, which seemed quite odd and a bit scary to say the least. Needless to say, it didn't go off. Yellowstone is one of the most unstable parts of our planet. It also has a geyser called Old Faithful, which erupts every 44 to 125 minutes, sending 4,000 gallons of boiling water to a height of 200ft for a duration of three to ten minutes. One awful story we heard was that of a young boy who fell into a boiling well and only his skeleton surfaced. Horrifying.

We will definitely return because we love America, even with Donald in charge. What a situation the world is in: a mad cheese-loving North Korean schoolboy with a strange hairstyle, a sneaky Russian and an arrogant billionaire, also with a very strange hairstyle and all three have fingers on buttons. Stop the world, I want to get off!

In the alligator's nest with the boat driver about to pounce.

Singapore, 1st Trip 2000

(One big launch and too much durian.)

I arrived late in the afternoon to find the agent's office closed. There was nothing to do but check into a hotel nearby and wait for directions to the shipyard where the ship/FPSO was located. They came the next day and I called in at the main gate security to obtain a pass and went to see the 'ship'. Ship is an understatement. As mentioned in earlier chapters, an FPSO (floating production storage and offloading) is basically a floating oil rig and this particular 'ship' stood out - not only as it was massive but it had a huge 'P' on the funnel for Pemex, the Mexican oil company I was working for. This must be the one!

The office was next door to the ship but nobody was expecting me so I was told to call back in the morning. I did and was left waiting in another office with an irate French guy who was in exactly the same boat. He calmed down when we discussed our situation, until a gormless portly Mexican guy in his mid thirties entered the office and introduced himself. To my consternation I found that he was my new boss and he wasn't expecting us either! As it was a new project and I was being paid well, I let my new French pal do the mouthing off. We returned to our hotels and called at the office next morning yet again to receive our instructions - which were to go with our assigned estate agent to find accommodation.

So far this job was easy peasy! We spent the whole day looking at places and, I must admit, they were impressive. I decided on an apartment on Orchard Road, right in the middle of town next to all the shops, restaurants and bars. The next day, our new boss found us desks and let us sort our own jobs out due to the fact (as I had predicted) that he didn't know what they were. This was largely OK except we had to write out daily reports to the Mexican's bosses and so far nothing had happened so making up 90% of the content was slightly awkward.

I was soon joined by my pal Greavsie, who brought along a friend called Gordon from a previous project. Things were eventually starting to take shape and best of all was the fact that our useless portly boss still didn't have a clue about the job and just played with his computer all day, only getting off his arse for the toilet or a coffee, before leaving early every day. The French engineer resigned but was soon replaced by another French engineer I'd worked with in Venezuela, who was also a useless fat pain in the arse. He had spent all his time in Venezuela arguing every day with - yes - yet another useless fat pain in the arse, a British engineer this time. (Were they breeding? They were certainly following me around.) It transpired that the new French engineer's sole contribution to the job was to say "Bonjour" and fall asleep at his desk. Greavsie and Gordon used to balance empty plastic coffee cups on his head but I thought: 'Best let sleeping frogs lie,' and wore earplugs to obliterate the snoring.

We were asked to meet a delegation from the oil company in a hotel in town. It was at the Four Seasons where President Bush used to stay, so security was a wee bit tight. The delegation looked like bandits from a spaghetti western, all swarthy, minus sombreros, bandanas and guns. They asked us things that were either impossible to answer or just plain ridiculous, the thinking being that questions were not for engineers to answer, especially as we were all new to the project. There were questions on how to cut costs and budgets. I did think they could save money by getting rid of the abundance of fat useless pains in the arses but, as the job was a doddle and well paid, plus the perks of finishing early every day, I kept my head down.

Burns Night crept up and I managed to get hold of some haggis and neeps, along with decent whisky and invited the two Gs to my place. Now, coming from Liverpool, I had never seen nor eaten a durian - a large spiky fruit that tastes really lovely and creamy but pongs like nothing on earth. The 'scent' is generally described as a blend of 'raw sewage, rotting flesh and smelly gym socks'. Durians are so stinky some taxi drivers refuse to carry them and it's even banned from public transport, hotels, even planes in some countries. Oh, isn't knowledge a wonderful thing? Of course Greavsie is one of those people who just seems to know these things. The Gs left me in the early hours to clear up. I couldn't smell anything at this point as whisky has a certain way of deadening one's senses, but the next day I awoke to The Pong. I couldn't for the life of me think what it was. Had something *died* in here and I'd been too pissed to notice? Then I checked the rubbish from the Burns bash and nearly passed out from the remains of Greavsie and Gordon's bloody durian. To add to this, I was expecting some good friends from Yorkshire who were stopping off on their way to Australia and now had to get the cleaners to get rid of this awful smell. When I got back from work it was still there, strong as ever! I searched the whole apartment. The sneaky bastard Greavsie had planted slices of the bastard fruit all over the place. The cleaners returned the next day to de-pong just in time for my guests' arrival but I could still smell it. In the end I cheated and took them for a drink at the world-famous Raffles Hotel to distract them. A successful weekend was managed but that bloody smell remained with me until I left the job.

The FPSO was launched from a wet dock with the usual pomp and crowd of people you'd never clapped eyes on in the building and start-up of it but who turned up in droves for the cork popping and backslapping. We then had to work on board, which was OK apart from the fact that our lazy pain-in-the-arse boss shared our office. The main problem with that, apart from his breathing, was his idiotic instructions. He had started to annoy me when we were in town as he'd hold meetings at my place at the weekends and go on and on about irrelevant items that he didn't have a clue about. I had mentioned stupidly when we first met that I'd been to Mexico and he would go on for hours about it to my detriment. As it was my apartment, I was left to sort things out when everybody else had left, which was always late.

The ship had moved out to the bay which meant that we couldn't get off and had to sleep on board. We were allocated a cabin and I was to share one with three French engineers but the crafty sods Greavsie and Gordon hung back and got an unused office as their cabin, all to themselves. The boxes of office files were now boxes of booze that we cunningly stocked up with for the coming dry voyage! Anyway, to cut a long story short, I ended up having a big bust-up with my boss and quit. So that was that. I went to see the captain to explain my predicament: I didn't want to go on the voyage but wanted to get off. I then told Greavsie and Gordon. Their reply was: "Can we have your booze?"

Initially, I had wanted to go on the voyage as it passed Krakatoa en route to Africa. Krakatoa, the famous volcano that erupted in 1883, is not a site one visits often but the voyage was over and it was arranged for me to jump/leave the ship at 3am due to the tides. I had to negotiate a rope ladder to climb down the side of this giant ship in darkness into a small Chinese boat to get to the landing stage. It was no easy task, I can tell you, as the bay was a bit choppy but I managed it.

Next day, I telephoned my agent to arrange my air ticket for the evening flight to Heathrow, then checked my bank account to make sure that my salary had been paid in. It wasn't the best way to end a contract. Life's not always 'a bowl of cherries'. Occasionally it's a bin full of durian.

Me with a fellow inspector, FPSO in background.

Me with my wife, Patricia, who flew out for the launch.

IRAN, 2ND TRIP
2001

(Where I give the 'thumbs-up' to oil-company politics and watch the Twin Towers come down.)

I wasn't exactly over the moon about returning to Iran, especially working for the French but it was a good position: head of the mechanical department. Oh and the recompenses were favourable too, which helped somewhat! After an overnight stay in Dubai, I flew to Kish Island in the Arabian Gulf for visa interrogation, where either you being a spy or them just plain scaring you is what it's all about. We eventually arrived in Asaluyeh on the mainland and it was hot. I was the only Brit in the accommodation as it was a French oil company and my boss was a short arse called Jacques Levois – I called him Leroy. The first morning meeting didn't go down too well, as my French wasn't as au fait as theirs but, as the project was in English, they had no choice in the matter. So Waterloo here we come again, put on your wellingtons! The main office consisted of the planning department and what little space remained was for the other departments. So planning had planned it well. As planning usually comes before anything else, I wondered why they still needed all this office space but empire builders are empire builders and they like to stay around.

My department consisted of supervisors, technicians and graduates. I read through my half-English, half-French handover notes to get up to speed on the project, then walked round the refinery with some of my supervisors. The plant was going through the final construction phase and would be followed by pre-commissioning, commissioning and then eventually start-up. The usual squad of vendor engineers were on site: Germans, Dutch and some Brits. I was eventually to relocate to the main workshops and office when it was ready, which was just as well before the planners planned to take over the world. It didn't take long to suss out who was going to be a pain in the arse and who was OK. The restaurant had its problems, like food! It was not covered; correction, it was covered but by flies. The general idea was to get in there before anybody else and get at the food below the liquid level as the fly shit was on the top, unless they were scuba-diving flies. It was that bad. I always thought that the French were gourmets but this restaurant lost its Michelin stars right away.

The evenings were spent either talking to my French colleagues, which was hard work and meant breathing a constant cloud of Gitanes and Gauloises cigarettes, thus taking time off your life every night, or in your room with book, laptop and DVDs. No competition there, then. Good job I'd brought enough books and DVDs with me. To my delight, they had tennis courts there which I could spend a few hours per week at. The day started with a one-to-one with Leroy to produce my plan of activities and provide him with some notes. Easy peasy. He would write it down as if he had planned it all himself, then go to his meeting with our project manager, who would go to his meeting with the Iranian refinery manager, who would go to the refinery director, who would inform the Iranian oil minister, who would inform the gods and so on. I would brief my supervisors on the day's work and then hold safety talks with everybody from the department, basically on how not to kill themselves until I am out of their country. Then it was to the morning meeting with Leroy and all the other departmental heads. A quick coffee and off to the operations meeting, where all the arguing, bitching and finger-pointing takes place. I have never ever worked on a project that was ahead of schedule, yet there has to be a scapegoat, somebody to blame, and this project was no exception. The masters of the universe, the planners, never quite get it right with their little lines, boxes and pies that never seem to tally with those little figures in the boxes above their masterplan called dates. Or maybe they are only one digit out, like a year! But if one promises the client the world, then he wants it now, as Mr Client is picking up the tab and thinking of all that black gold and profits and Rolls-Royces or, in Iran's case, war fund. Of course they want their returns ASAP, so you see, this is why we get it in the neck at the sharp end. This is oil-company politics, strategy and policies the world over, making rich people richer!

Driving from the office one night, I gave one of the Iranian supervisors a lift. As we passed the main gate, I gave the usual thumbs-up sign.
"No, Mr Ken."
"No what?"
"This mean very bad thing."
"What, the thumbs-up?"
"Means same as American finger sign in Iran, Mr Ken."
'Oh shit.' I have been driving passed armed guards every day, morning, noon and night, giving them the finger. Talk about learning the hard way. I suppose it could have been harder, like collecting a bullet between the eyes. The next day, I drove passed the guards with a smile and a wave. 'Phew.' Cultures, hey?

I couldn't get into the main office. 'What's up now?' Always something. The door was quickly unbolted.
"Why is the bloody door locked?"
"Rioting has just broken out."
"Where? Over what?"

"Some of the Iranians have been laid off and are taking it out on the Koreans."

Koreans, being the main contractor, had laid off some local Iranian sub-contractors which didn't quite please them at all and I wasn't a bit surprised. So here we go again. We remained locked in the office all day without any food. Leroy called me into his office.

"I need you to drive down to the refinery in Bandar Abbas (Southern Iran) with the head of planning, to interview some Iranian engineers. We need them on this plant as we have a shortage here."

"How many do you require?"

"As many as you can get."

We set off at dawn the next day. It took us nine hours to get there with just one pit stop and we had to stay overnight in an Iranian half-star hotel. At least the beds were clean and the cockroaches didn't bite! We arrived at the refinery the next day at 9am and, of course, the security weren't expecting us as they'd not been informed of our arrival. After an altercation which took about an hour, we were introduced by the refinery personnel (more like seduced, as they do like to kiss and cuddle). Eventually, we settled ourselves in an office and started interviewing the Iranian engineers. Most seemed very nervous and I couldn't understand why as the interview wasn't too stringent. Lunchtime came and we were taken to the refinery restaurant, where we all sat round this huge table. In walked the directors and managers of the refinery, along with the ubiquitous ayatollahs, in all their glory. The meal was a banquet; these religious bods don't miss a trick! After lunch, I walked back with a young student ayatollah and he broached the subjects of religion and current world problems. Very delicate topics to discuss with a soon-to-be ayatollah in Iran. I was, of course, very careful about what I said to him but it was a fascinating conversation. They have their eye on the ball, i.e. the world! He knew a great deal about the West and its politics and world affairs in general; he just seemed a normal guy - polite, well mannered and obviously well read.

What a day. We must have interviewed about forty engineers and technicians and chose just six. Not a good strike rate but one can only pee with the tackle one has got, so to speak! Then it was back to Iranian Towers and Basil Ayatollah Fawlty. The planning department had just gone up in my estimation as the planning manager managed to acquire some cans of beer, which was very welcome indeed. They can have the whole office now. After all the interviewing and that horrid two-day drive, the refinery wouldn't release any of their best engineers and why should they? I think their masterplan was for us to take their deadwood off them. It had been a complete waste of time and effort.

The main workshop was almost complete so I moved into the office and away from planners and Frenchmen. It was just me and the Iranians, who I got on well with. The workshop manager, who was in his late sixties, I soon christened Mr Fish, as he'd put a notice on the noticeboard using a sketch which looked like a fish. I returned from a field break with one of those fish plaques that sing and dance when you clap your hands. That went down a treat and broke the ice. Mr Fish could not live that down as all his staff called him it now. Every time I called into his office for a coffee and a chat, I'd clap and get Fishy going. We all worked well together, and I liked the Iranian people. It's the same the world over: ordinary people are fine; it's just the people in power that cause the problems. Just think, we could solve all the world's problems with a clap and a little fish. If only that were true! Anyway, we are halfway there as we already have dead fish running things but they need a battery up their arses and instead of a clap they need a slap.

I walked into the fly-house, sorry, restaurant. "Hey, Ken, have you heard the news?" It was my Cloggy mate. "The World

Trade Center in New York has been knocked down, the twin towers, both of them!"
"What, you serious?"
"Go into the TV room."
I shot into the smoke-filled room, where most of the French were gathered round the TV. I couldn't believe what I was seeing as the towers came crashing down.
"Jesus Christ, what the fuck is going on?"
"Two planes have just crashed into the towers, monsieur."
I was stunned, shocked and felt sick. I had been in those towers only weeks before on vacation and remembered asking a woman to take my photo as I stood in front of the radio mast that was now collapsing and falling through all the floors. The towers were so huge and awe-inspiring, you found yourself looking down on the Empire State Building with the Statue of Liberty like a toy soldier. I had taken photos around the base of the towers looking up; they were that awesome. Then a third plane crashed into the Pentagon, where I had visited also. What a horrid coincidence. I had been to Boston and remembered watching planes taking off and landing from the very airport that the hijackers flew from while having a beer in a bar on the quay. I had visited the Pentagon and been up to the top of the World Trade Center. I watched repeat after repeat in total disbelief. 'The evil bastards.'

I returned to the restaurant. The cook was wearing a black bin bag as an apron – somebody called him Osama Bin-Liner but no one laughed. I couldn't eat and went to my room. After that, the end of the assignment couldn't come quick enough. On the day of departure, we drove to the airport which was just a five-minute drive from our base, only to discover that it was closed.
"It's closed, mister."
"What do you mean the airport is closed, how can the fucking airport be closed?"

I was in a foul mood after all the bad news! Rumour had it that a certain local official hadn't received his monthly backhander so the dear chap had found it necessary to close down the airport. 'Great.' The key in my back wound up a few extra notches.

"So what do we do now?"
"Mister, go bus to other airport."
"Where is other airport, how far?"
"Maybe two three hours over mountain pass."

'Oh shit.' I had been on these roads before and one slip meant certain death as you are only a foot or two from the edge of a shear drop. The local Iranian bus contractor, I would assume, had never even heard of an MOT certificate, let alone a service, and this was hardly the time and place to test the brakes.

"No problem, mister."
"Are you coming too?"
"No, mister." That said it all! "Mister?"
"What?"
"Be careful."
"Yes, I know the road is dangerous."
"No mister, al-Qaeda maybe in mountains."
Well, that boosted my confidence no end.

After four hours of driving, trying not to look down and hoping that the local terrorists didn't show up, we arrived at the other airport. In the security queue, I was behind a bad-tempered Frenchman who was getting annoyed at having to empty his bag.

"Oh just shut up, will you? This is hardly the time and place and certainly not the people to piss off!" It was my turn. The security guard had hold of my wrist heart monitor that I tend to travel with and was shaking it – you know, the ones that pump up and make a noise. I was thinking: 'Oh please don't press the screen...'

"What is?"

"It's a diving computer." He handed it back.
"Shukran,' I said. Oh shit that's Arabic - he speaks Farsi!

After about an hour's wait, the oil company's plane arrived and we started to go through the gate. I was last through.
"Hey, mister." It was two armed guards. 'Oh fuck!'
"Yes?"
"You British?"
"Yes."
"Problem?"
"No, mister, no problem. We are students from Tehran university and we want learn speak English."
What, now?! "You kind of caught me at a very inconvenient moment in time."
"We ask you questions."
Bollocks.

They asked me questions about Virginia Woolf and George Orwell's Animal Farm. Fortunately I'd read Animal Farm.
"Mister, what meaning high and low?" They couldn't understand the difference between high and low (in height and depth) and high and low (in sound). So I started 'orrrrrr-ing' for low and 'eeeeee-ing' for high! By this time, the French were witnessing a crazy Englishman, who had been arrested by two armed guards, going through a theatrical repertoire of strange noises.
"Regardez cet Anglais, qu'il est tou, oui, Napoleon ovait raison." (Translation: "See the fucking crazy Englishman. Napoleon was right about them!") I could see the pilot mouthing: "Nous sommes prets a partir, vous s'il vous plait depechez-vous, et obtennez votre cul sur le plan." (Roughly translated in Scouse: "We are about to effing leave, will you get your arse on the plane?")
"Well, guys," I told the students, "I got to go, plane is waiting and pilot he get angry but I will sure come back and help you out on my next trip!"

Next, Kish interrogation island and a two-hour wait. There are no bars or shops, just an agonising wait, so near sanity. What a wonderful sight Dubai is after all this hassle. I cannot understand why the Iranians in power behave like this. It's very scary when you think they are so close to having the ultimate weapons! At the hotel in Dubai, I climbed into a nice relaxing foam bath clutching a cold beer, radio on. The key in my back was unwinding when I heard on the news that the Beatle George Harrison had passed away. So now there are two.

Me and my Iranian Range Rover.

Korea, 2nd Trip 2003

(Collapsing cooling towers, Barney's barneys and a bitter end.)

I flew into London Heathrow for a meeting at Britannic House, BP's headquarters, before flying to Korea to start a new project. Heading for the exit, I noticed a guy dressed like Lady Penelope's chauffeur, Parker, who in turn reminded me of Noel Gallagher. He had a noticeboard with my name on it. 'Bloody hell, this is impressive,' I thought. Parker took my luggage and we walked through arrivals to the car. I couldn't help notice people looking to see who I was! After the meeting, my new boss took me to the BP restaurant for lunch. "What wine would you like?" This is not at all a bad start to a project. After lunch, Parker drove me to Heathrow and I was soon on my way to Korea. I arrived full of jet lag and checked into the Hilton. This was getting better all the time, as the Beatles song goes! A note had been pushed under my door. It was from the project manager, welcoming me to Korea and the new project and inviting me to meet up for drinks in the lounge bar at seven.

I set my alarm and tried to grab a couple of hours sleep and awoke, feeling dreadful as one does with jet lag. I showered, dressed and went to the lounge bar to meet the PM. I was a little early, so ordered myself a beer. He showed up and we shook hands, ordered more drinks and then he hit me with a bombshell. "Now, if there is any fighting on my project, you will be gone on the next plane."

"Pardon." I was bemused and yet taken aback at this comment. "Sorry, didn't quite catch you there." What gave him cause to make this statement, I was soon to find out. A Scottish welding inspector, no less, was playing pool with an American welding inspector and, after a night on the juice, decided to bet on the game. Of course the wee Jockey soon found himself down on his Hogmanay funds. Well, Yanks do play pool quite a lot. The nasty wee Jockey took this as his cue to use a cue to whack the poor fellow over the head.

"Oh, I see. We're all tarred with the same brush!"

"No, no, it's just that I cannot afford to let this happen again."

"Well, I assure you that I hate playing pool, I don't gamble, nor am I a drunken Jock."

The following day, I was driven to the site which was four hours from Seoul through beautiful scenery. The hotel we were all staying at was in the countryside, surrounded by farms and close to a small village. It seemed very antiseptic, as it had those dreadful white strip lights that I can't stand and stainless-steel banisters. Every time you touched them you'd get a bolt of static electricity which made your hair stand on end. The dry air in this region didn't help. The room was quite cozy, with a good view of the farmlands and the farmers with their strange little tractors that looked like lawnmowers. Later that evening, I met the crowd and soon realised the situation I was now in, being the only non-staff BP guy on the project. As I spoke to each staff member in turn, I was cautioned not to get involved with such and such a person. The BP staff had worked with each other before and all bore a grudge with one another. "Don't mix with him, he's a c*nt!" (or in BP staff speak, "cant") and so on. 'Oh, this should be fun,' I thought.

We each had a car and a driver as we were not allowed to drive on the roads in Korea, Koreans not being the safest of drivers. So each morning, your driver would be waiting with a warmed-up car, but as soon as you got in you wanted to get out as it would reek of kimchi, a Korean national dish made out of chilies, radish and Chinese cabbage. You'd see the locals sitting in the village square with all this on a sheet of polythene – they'd be up to their armpits mixing it – then would bury it for maybe a year or so until it fermented and became a delicacy. There are kimchi cellars like wine cellars in Korea, run by professionals. Personally, I couldn't stomach the stuff. It's supposed to be one of the world's healthiest foods, rich in vitamins, but some reports say it can cause gastric cancer! Many Koreans like to eat dog, so if you were having a meal and found a medallion in your mouth, beware. Sometimes you'd see a local on a motorbike, riding along with a dog in his basket: that was their Sunday lunch. Much prefer roast beef and veg. The drive to the refinery was rather an exasperating journey. When we arrived, the driver would get out and go into the driver's room until midday; you'd then have the car to use on site and pick him up to do the reverse. I was the only expat in our office, the others were Koreans, which was OK but it was tiring having to speak pidgin English and answer their questions most of the day. When they spoke to you, they would stand about a foot from your face and shout, with their kimchi breath making your eyes water.

There was plenty of activity on site and many vendor engineers to liaise with, so being very busy made the days fly by. In the main office, there was a mixture of English, Scottish, Americans, Dutch and Australians. I was quite pleased with the set-up and it was very interesting. The refinery was huge, with lots of jobs going on all over the place, and it was on the edge of the Yellow River, China's second longest river. After a month on site, BP would allow you to spend a long weekend in Seoul, midday Friday until midday Monday. Staying at the

Hilton, all paid for, with some spending money as well, was a good deal. On one occasion, John, a BP guy I'd befriended, and I decided to venture away from Seoul and try another place for a change. We checked into a small Korean hotel. The manager showed us to our rooms. "John, we can't stay here, there isn't a bed." It was in the wardrobe, a roll. The tariff was way below the Hilton's, so we were quids-in. Anyway, after a few drinks I would not notice. But the next day I did, as my body was stiff and aching. We spotted a sauna and thought it a good idea to relax our knackered bodies after sleeping on the floor. The place was full of naked men which didn't appeal to me. They were washing each other and a male masseur was doing the biz on a naked bloke. "Shit, I don't like the look of this place." It wasn't a homosexual place but I didn't like being naked in front of men. Women, yes! I turned down the offer of a massage and was glad to get out of the place! It served us right for going cheap.

The problem with Seoul was that it was full of young drunken American conscripts who acted like a bunch of twelve-year-olds. There would be fights breaking out on nearly every street corner. The Korean War all over again. Americans against Koreans, bashing each other's heads with planks of wood and whatever they could lay their hands on, not that it would have made any difference, wood on wood! The shopping was good, especially at the most famous Itaewon Street, full of shops and markets. I used to return to Korea after home leave and only bring my briefcase and a small bag, not even a suitcase, buy one there and fill it with goodies to take back to England. Christmas came and we all went home. On our return to Korea, we all met up in the business lounge at Heathrow. Most of the guys seemed really down, the reason being they hadn't declared anything coming through customs at Heathrow and had everything confiscated, followed by a hefty fine! As I came through, I could make out the customs officers with their bright white shirts and black ties, looking very intently from behind what was supposed to be a one-way window. So I declared my items (not to say that I wouldn't have anyway!) and they charged me a paltry £15. So I was on a winner but of course never mentioned it as I didn't want to be the smart arse. I did have a laugh to myself, though!

One day, driving past the cooling towers, I stopped to have a look at how work was progressing on the pumps inside the cooling-tower pit. The pit didn't seem deep enough to me for these types of pumps and would interfere with the NPSH (net positive suction head). I was concerned that it would create a vortex and suck air into the pumps, making them cavitate (when fluid vaporises and causes implosions, causing serious damage). I reported this to the Koreans. Their answer was to install a vortex breaker, which I doubted would solve the problem. I asked them to make a small scale model, the type I used in the laboratories at the university in Quito but they didn't go for the idea. A week later they were working in the pit to make it deeper. Now, I don't know if they took my advice on the scale model but I reckon that they did, as they were now excavating like mad.

I had been on leave and had just arrived at the hotel. As it was near lunchtime, I went into the dining room and kept bumping into people that kept telling me that the construction and commissioning managers were looking for me. I drove to the refinery wondering what they wanted me for. I was soon to find out. A cooling tower is a very large structure and there were three of them. As I drove past I couldn't see them. 'What the?!' I screeched to a halt. Where are the fucking cooling towers? I got out of the car and walked through all the rubble! Next moment two cars drove up and out popped the construction and commissioning managers. My mind was working overtime, thinking of what I did on the towers before I went on leave, which was to check the cooling-tower fan alignments. As they walked towards me they both said in unison: "We need a full report on this first thing in the morning". 'Oh shit.' What went wrong here? The huge blades of the fans were melted and stuck to their gearboxes. It was as if a bomb had hit them. I spotted a sly grin on the construction manager's face. The story was: inside the towers there are lots of slatted wood on the walls and a welder had been welding some brackets without a firewatch and 'whoosh!' - the tower turned into a towering inferno, followed by the others. The whole lot went up in flames like paper with a gigantic flue overhead. This, of course, destroyed the concrete, resulting with all three towers having to be brought down and rebuilt. Not quite sure what happened to the welder! But I'd thought I was in the merde for sure.

It would not be unusual to see a ceremony taking place when something had been commissioned. The Korean engineers would have a table set out with beer and peanuts etc, surrounded by flags and all sorts of buntings. Only problem was, they sometimes jumped the gun and held a celebration before anything had been tested and when you mentioned this to them it would go down like a damp squib. The Japanese vendor engineers invited me out for drinks at a local bar one night. One of them was sitting next to a Korean women, fondling her breasts whilst eating peanuts, drinking beer and holding a conversation with me as if she wasn't there, treating her like a plaything. I suppose to him, she was: he 'paid and played', so to speak. She just smiled and ate peanuts as he fondled away! One compressor engineer from UK found himself in love with a local lady and just disappeared, never to be seen again. We were amazed, as he was no George Clooney, more George Formby, and she was a looker. She must have liked his banjo. And there was Barney. Barney Luckenbach, an ex-LA cop who was built like a barn too. Barney was some character. BP had a policy that we were not supposed to give vendor engineers a lift, which I thought was a bit over the top. With having so many of them on site at the same time, it was difficult to get to see all of them in the day. I had to get information from them for our daily progress meetings so I ended up giving Barney a lift back to the hotel. At first this seemed a good idea as I could gather all his information on the way, but Barney couldn't pass a bar. We had to drive up this little dirt track from the refinery which went right past one. "Just for an hour," he'd say. "Yeah sure, Barney, an hour." With that, we would tell the driver to come back in an hour.

The hour soon passed as he was really good company. Of course we'd send the driver away again and this would be repeated on the hour until the bar closed. One night, we had to walk back to the hotel and found it closed and had to knock the night-duty receptionist up to let us in. We found a small bar in the village with no name, bit Clint Eastwood, but it had a sign with a saxophone on it, so it was christened 'The Saxophone Bar'. One Saturday night, I met Barney in there and sure enough he had a woman in tow, a local. After a few drinks, he went to the toilet and was gone some time when I heard a commotion from the next room - shouting and bottles being smashed (not unusual in these places). I went to look for Barney and, as I approached the other bar, saw bodies flying through the air. There he was, in the thick of it, chinning all these Koreans. Somehow I managed to drag him away without being chinned myself.
"What the hell happened back there?"
"One of them Korean bastards called my sweetie a whore!"
"Well, she is, isn't she?"
"I didn't like the way he said it!"

Barney would disappear for days, off on another job, then I'd be sat at my desk and the door would open and a huge shadow would be cast on the floor. 'Oh no, not Barney again.' I'd just got over the last hangover. He'd get his camera out and show the Korean engineers his latest holiday snaps, mostly of naked women - what else?! His camera was a new digital type that displayed the date the photos were taken. All these photos of different women bore the same date - the Koreans were in stitches.

There was a small grocery shop just round the corner from our hotel: 'Props. Mr & Mrs Hong'. As the bar in our hotel was too antiseptic with its stainless steel and strip lighting, we needed a more friendly atmosphere. One of the Americans spotted this shop that sold maekju (beer). The place was like something out of M*A*S*H, old and decrepit and leaked when it rained and when it rained, it rained! It was perfect and we soon christened it Hong's Bar. I made a wooden sign with Hong's Bar on it and hung it over our enlisted 'bar' made of stacked empty beer crates. When various personnel visited, they'd put their business card on it. Of course Mr and Mrs Hong loved that and were so proud. The guys would see each other during the day and say, "See you in Hong's!" We'd sit on empty beer crates and out would come the 'when I was in' stories. It got hilarious, especially with Barney. His calling cards were fake Rolex watches, which he'd bring back by the dozen from his trips. When a female customer came in, he would dangle the fake Rolex watch and say: "Rolex watch, jiggy, jiggy." They'd laugh and run out. Good job Mrs Hong didn't understand what he meant as she was a very religious woman and would have thrown him out, despite his size. She gave Mr Hong some stick if she ever saw him having a beer with us.

One Saturday night, my mate John and I called in to Hong's and were having a beer when two local farmers came in and joined us. Neither could understand the other, so we started drinking soju (distilled spirit), introduced to Korea by the Mongols circa 1300 when they were doing their usual bit of marauding! They had marauded in Persia and extricated the recipe from the Persians. The ABV ranges from 18% to 45%. John went outside to the loo and was gone a while. Not another one! The only problem with Hong's bar was the lack of a loo. They used a large hole in the ground and one had to be careful as there were no seating arrangements or lights. With the copious amounts of jungle juice John had consumed, I thought I'd better check and see if he had fallen into the stink pit! If he had, God knows how I'd have retrieved him. 'Sorry, mate, got to go!' I went outside. It was bucketing down and John was holding on to a downspout, which had broken away (courtesy of John) from the above gutter and was spewing gallons of water all over him. We really enjoyed our soju night with the farmers.

Before the coffee shops in the West kicked off – Costa, Starbucks etc - Korea had them and they were quite quaint, cosy and warm. What amazed me were the women who ran them. Although the term 'coffee shop' was an obvious front for devious business, it was the amount of coffee they consumed in one day entertaining customers and potential customers. They flitted from one table to another and had a coffee or two at every one! They began to look the colour of coffee. One Saturday, I went for a haircut and walked in to find it very crowded. There were at least a dozen barber's chairs and each one had a body in it. But the odd thing was, there was only one barber tending to one head. The other guys appeared to be asleep and each was covered with a blanket! 'How strange,' I thought, 'blow this, I'll come back midweek when it's not so crowded.' On the Monday in work, I mentioned this to some of the guys who'd been in Korea for some time!
    "Did it have one or two revolving signs outside?"
    "Not sure, why?"
    "Because the two sign ones are the haircut/relief shops."
    "What?"
    "When you have had a haircut, they put a blanket over you and dainty feet will come scurrying over, put a hand inside the blanket, unzip a banana, fit condom and hey presto sweet relief!"

'What's that prick up to?' I was coming out of the hotel one night to find the construction manager hiding behind a bush! Walking along in front was one of the female office staff. He was stalking her, going from bush to bush following her, and he nearly died when he saw me. That put an end to it. I never got on with this guy; he was a devious swine to put it mildly and would try to catch you out and drop you in it as much as he could. We celebrated Burns Night with a formal dinner at the hotel. He and his fellow jock (remember the cue-wielding welding inspector?) were dressed in their kilts, sporrans and sgian dubhs. It looked like a United Nations dinner – well, I suppose it was in a way, as there were Japanese, Dutch, Americans, Canadians, Koreans and Australians. Of course the Jocks had rehearsed their speeches and performed with aplomb. Then he asked me to stand up and give a song. Well, I am not the best of singers but that didn't matter: I didn't know all the words to any song. I tried a rendition of 'In My Liverpool Home' but never made it and sounded a real arse. I caught a glimpse of the gruesome twosome smirking. They had that 'we got you' look about them, the scheming bastards.

The following day, I was called into the project manager's office to be informed that my father had passed away. So I was on the next flight home, first class, courtesy of BP. To top it all, I was reassigned to guess where? Bloody Scotland.

ANGOLA
2004

(Where I am attacked by fangs and claws and get an unexpected piggyback from a whale.)

My mate Guy 'Greavsie' and I flew out to Angola to join our new project for the Angolan oil company Sonangol. As we couldn't fly direct for reasons beyond our control, we had a stopover in Johannesburg. It wasn't a problem as we weren't in any hurry and it gave us just enough time to have lunch and a couple of drinks prior to boarding. It was nighttime when we arrived: an oil-company driver picked us up and took us to our apartment in downtown Luanda, the capital. The first impression was not a good one. Although it was on a main road, it was dark and seedy and I found that I was crunching things as I was climbing the steps to our floor. I soon realised that the crunching underfoot were cockroaches, which kind of left one with a somewhat sickening feeling, not a bit welcoming and heartwarming. The apartment was clean and adequate inside, considering.

The next day, as the cupboard was bare, we got the driver to take us to a decent place to eat. He took us to a place by the ocean called the Illya, which was a kind of boat landing quay/peninsula, with many Western-type restaurants. We settled on a Portuguese one, where the ocean almost came under your table when the tide was in. After brunch, we made our way to the oil company's office. You could see the Russian influence here, with hammer and sickles everywhere. Angola's flag is a hammer and sickle too, mixed with red and black. You could also see the Portuguese influence, with many restaurants and bars, as it was once one of their colonies. We met up with the flock, the usual bunch of worldly oil men, the 'Been around, done that, got the T-shirt' types that you meet on these overseas contracts.

The top guy was, as expected, an American and our immediate boss was French. The office was adequate with plenty of coffee on tap. The job was overseeing repairs and recommissioning some of their offshore oil rigs, which had been neglected. Some were in a dangerous state and full of H2S, hydrogen sulphide - a very lethal and toxic gas, particularly dangerous as you cannot see it but if you get a full blast of it, it will kill you in seconds. So not pleasant working conditions but at least it got you away from the office! The heliport in the mornings was not a treat, as you had sweaty oil labourers and mosquitoes who had not had their breakfast. Before we were allowed offshore, we had to take a survival course. The drive there was no picnic as we passed through some real downtrodden and very poor areas. An American engineer opened his window of our car as we were driving along to take a photograph - bad decision. He no sooner sat down when an arm came through the now-open window and tried to take his camera and it wasn't a Brownie box type, so photoshoots were out after that unless someone was watching your back, front and sides!

It was the usual survival course (see earlier chapter) but at least it was outside and warm so a bit different from Aberdeen. We had to jump off a high diving board with our life jackets on, swim over to a life raft and climb on board. As I was hauling myself up, what should greet me was a cloud of half-conscious mosquitoes which I had to splash away with water before I dared climb in. The rest of the course was routine; the main difference between this one and a UK course was the temperature of the water. In the North Sea, you are up against the cold and hyperthermia and after about two minutes Mr Reaper calls on you. In Africa, it's very possibly sharks and other nasties nipping at one's part of the anatomy that sits on chairs!

Our day started with Hungry Horace Greavsie opening the bloody window of our apartment to feed the birds and, of course, thus allowing the mini Draculas in to feed on us during the night. But my protestations were adhered to at least, as I didn't quite like the idea of catching malaria - it was bad enough having to take the prophylactics to keep the buggers at bay. The traffic was horrendous at all times in Luanda. After the morning meeting, we would head back to our office, check emails, plan work ahead and make arrangements to go offshore, and at lunchtime we usually went to an Italian restaurant which was close to the office. Whenever I paid a visit offshore, I left my passport locked away in the apartment and took a photocopy with me, which had always been accepted by Angolan immigration. This particular day, I was in a hurry to get back to watch England who were playing Portugal in a European match. On arriving at the heliport in Luanda after my trip offshore, the immigration officials wouldn't allow me in. I couldn't come in and I couldn't go back out again without my real passport. So I found myself in a rather tricky situation – kind of stuck in the middle with me! I was more concerned about the time, as it was getting near kick-off. When in this situation in Africa, you bribe your way out, which I did with a $20 note. I could have had anything stamped for that but the photocopied passport did its job and I was through, and true to form, bloody England lost!

One morning, we were heading to a malfunctioning satellite platform to check it out. It was about five am and still dark, and most of the personnel were asleep, when I noticed a large shape loom out of the water. It had to be a whale. It dived under our boat and I waited... and of course it started to surface and we started to fly. Talk about frightening – well, it was for me as I was the only one awake but was soon joined by everybody else. There were several cries of "What the ^$%@! was that?"

"Oh just a playful whale!"

Life was not too bad in Luanda. At the weekends when not working, you could take a trip down to the beach. Hungry Horace Guy managed, unbeknown to me, to get a load of lobsters from a local fisherman and had put them in the fridge and of course yours truly opens fridge to get a nice cold beer and was confronted by claws. The fridge was full of them. I was guilty of being stupid as I had almost blown up the cooker. I turned on the gas on what I thought was an electric ceramic hob like like the one I have at home, when all of a sudden it shattered into hundreds of pieces. Fortunately, I was walking away from it. Guy wasn't too pleased as it delayed our lobster thermidor - it was almost lobster thermaldor! But we survived to tell another tale. I bumped into an old Portuguese friend, Ronaldo, in Luanda from way back to North Sea days, who had a holiday weekend villa on a small island off the coast of Angola. He would pick us up in his boat and take us to have a barbecue and a few beers there, so weekends could be good.

However, life wasn't too easy in Angola either. In our office we had a tight-fisted Jock, which should read 'Joke'. Tom was his name. Him and Greavsie were fitness fanatics, so I joined in trying to keep the capital D away from my belly but alas, it didn't work too much. Tom was one of the electrical engineers and spent most of his time at his desk designing a house that he was having built in South Africa. Greavsie, renowned for loving food as mentioned, would find these little out-of-the-way restaurants and one night the three of us went to one of his favourite places, which I admit was quite good. As we waited at the table for the waiter, the tight-fisted Jock went over to a recently vacated table and took a fist full of chips that a couple had left. I mean, how uncouth is that?! This is Africa remember, God knows what had landed on them as they'd been sitting there but that was Tom. Funny thing was, although I'd not done or said anything bad to him, he just didn't like me at all and, by coincidence, the feeling was reciprocated. Funny old world, isn't it?

Then one day, I was called into one of the oil company's manager's office, regarding one of their floating oil rigs that had suddenly stopped floating and almost sunk in the harbour whilst being towed from Texas to Angola. I was tasked to find out, if I could, what had caused it and survey the damage. Fortunately, it had only partly sunk and had been towed into a dry dock, no need for scuba gear. So it was goodbye, Angola, and off to the Yellow Rose State...

*The road to ESSA Training Centre
(to do your survival course in Angola).*

On the road to ESSA. There were people living in that car.

Boat back from the local bar, Angola.

Korea & Kuala Lumpur, Singapore, Malacca, Langkawi, Vietnam, Cambodia

2006

(Where I do a bit of sightseeing...)

It was back to Korea on a new-build FPSO in wintertime. The flight was direct, business class on a Korean Air 747, so I was pretty refreshed which is unusual after a long haul. I got to our hotel only to find that it was full and I was booked into another one down the road. I first met my new colleagues for a drink and to get up to speed on the project. The hotel seemed OK from the outside but the reception was odd, just a small window, and as I pressed the bell, a person whom I couldn't see came for my details. The blinds covered his face and as I was in total darkness, he could hardly see me at all. This was most peculiar but, as it was late and I had an early day ahead of me, I thought nothing of it. I was to soon learn that the place was known as a 'Honey Hotel', where 'courting couples' go for a bit of privacy away from their families. After three days trying not to think about my mattress, I got into our proper hotel with a proper reception and met my new colleagues. One was an Irish engineer who was very witty and whom I gelled with immediately, the other was a Polish engineer, a serious type of guy whom I got on with after some time. We were joined by another engineer that had been in the Royal Navy on nuclear submarines and who seemed to think he was a cut above everybody else. We got on but only, may I add, professionally.

After three uneventful months working on the FPSO, it was departure time. We were on our way to Africa, being towed by two ocean-going tugs, and most of the office staff waved us off from the quayside. In Kuala Lumpur, we docked to change crew and were joined by some of the American crew's wives. I spent a few days sightseeing there, mostly visiting temples. One in particular had monkeys lining the walls that tried to snatch things from you as you climbed the steps to get inside. I also managed to get two tickets for the Malaysian Grand Prix as my wife was flying out to join me in a bit of travelling. Before she arrived, I agreed to go with some of the oil company's staff on a coach trip to a place called Malacca, 120 km from Kuala Lumpur. Malacca is the unofficial capitol of Malaysia, a beautiful historical place on the Malay peninsula.

Personally, I hate sightseeing coach trips and this was a typical follow-the-leader tour around the shops, markets and churches, with the obligatory group instruction talk from the infant-school teacher – sorry, guide. I immediately wanted to be away from the Americans in their polyester pants and the other nitwits who love this sort of thing, especially as I'd had a few beers the night before and needed coffee. Of course, after an hour in the coffee shop, nitwit me missed the time to return to the coach which had gone without me. Not to worry, first have a cold beer and work out what to do. My plan was to hire one of those bikes with a passenger seat at the side and get the driver/peddler to ride around and see if the coach was outside a cafe, church or shop. After a few circuits around Chinatown, I dispatched to make further plans. First plan: another cold beer. Then I spotted a bus with a Kuala Lumpur sign. In my rush for the bus station, I ordered another cold beer but immediately cancelled it as I realised I didn't know what "stop the bus I need a pee" in Malaysian bahasa was and it was going to be an hour or two at least on this bus. I was nearing 59 years old, not 29, and the idea of having to hold on for that duration did not fill me with glee. I finally arrived back in Kuala Lumpur, only to find out that the oil company had sent out a search party to look for me.

My wife flew out to join me and we stayed in Kuala Lumpur for a week, managed to get to the Grand Prix then it was on to Langkawi. We decided to tour around this area of the world as when else do you get the opportunity in life to see such places? Langkawi was gorgeous and our accommodation was on the beach itself. One sobering thing was that you could see the tidal mark where the tsunami had hit, a constant reminder of the power of water. We did a lot of beach things in and out of the water and I found a go-kart race track nearby. There were some guys from UK staying at our hotel, so every day we would meet up and race while my wife lounged on the beach. Many happy hours jet-skiing, sampling various tropical cocktails and dining outside in the evenings. Definitely a place I would revisit....

After a few days, we booked a trip to Singapore with a stay at the world-famous Raffles Hotel to treat ourselves. When we arrived, the receptionist had got our name wrong so we weren't on the reservation list so I grabbed my briefcase from the taxi and pulled out a copy of our faxed reservation. This seemed to embarrass the manager who gave us an upgrade to the Presidential Suite for the inconvenience! We were shown to our suite by a guy in a white suit, who introduced himself as our butler and was on call for us 24/7. We quickly put him to use by ordering our first of many Singapore slings outside on the veranda. He discreetly informed us that the Clintons had stayed in this suite the previous weekend so we found ourselves sleeping in the same bed as Hillary and Bill (a week apart,

I must stress). The dining room had a huge table used by President Bush Snr and his cabinet for meetings when he had stayed in the same suite. I sat on each chair to make sure that I had sat in the President's chair and made pretend phone calls on the desk phone in our lounge that the President must have used. I tried to put it out of mind when on the loo. There are limits.

Me on Clinton's phone, Raffles Hotel Presidential Suite

Before dinner, we ventured into the famous Long Bar with its floor covered with discarded peanut shells due to the custom of staff dishing bowlfuls of peanuts on all tables and clientele chucking them on the floor until the whole place was carpeted. The hotel has its own shops which are of course all upmarket Tiffanys and fancy clothes shops. It's also home to the Writers Bar, a tribute to famous writers who have stayed in the hotel over the years and where you can relax and have a bottle or two of champagne. We spent three nights there in all, before deciding to continue to Vietnam. The manager, who was still apologising, booked our hotel reservation in Vietnam, in another top hotel and where I awoke the next day wondering which country we were now in as I'd done quite a bit of travelling. This was a hazard of my job and something I occasionally did at home in Liverpool! It didn't take long to figure out where I was as I opened the curtain and saw a million people riding small Honda 50 motorbikes, wearing straw paddy-field hats. This had to be Saigon. It was hard to imagine a war had taken place here as, apart from the traffic, it was peaceful. Crossing the road was crazy, however, as the small motorbikes would be criss-crossing each other without stopping or slowing down and nearly all were carrying more than one passenger, often the wife and kids and usually livestock tied to the back seat. It was not unusual to see a bike loaded up with the weekend roast, i.e. a live pig strapped at the rear. We just stood gawping in amazement that there were no crashes, not even jostling for positions, swearing or finger signs given; they just got on with it! A Piccadilly Circus in downtown Saigon.

We visited and went down into the notorious Viet Cong tunnels which were extremely confined and stagnant: you could just imagine an American flamethrower heading for you! I couldn't wait to get out. There were many holes in the ground where the Viet Cong, when being pursued by the Americans, could just open the lid, drop down and completely disappear. What was sickening were some of the booby-traps that had been laid for the American soldiers. There were hidden holes in the ground that an unsuspecting soldier could fall into and land on sharpened bamboo stakes called punji sticks, real evil traps that would cause untold agony and or/ death. Another frightening trap was when an unsuspecting soldier opened a door to a house, a second door with sharpened bamboo spikes protruding would spring at them! This is now banned by the Geneva Convention. War is not a very nice thing, is it? Never has been, never will be. We visited a rifle range and had a go firing AK-47s, which were not at all accurate, as you could aim at something 100 metres in the opposite direction and still hit the target. Even so, I'd not like to be looking at the business end of one of those. With my luck, the bloody thing would be the one that had been calibrated correctly.

So it was on to Phnom Penh, Cambodia. Not a place I had longed to visit but it was only a short flight away, so the opportunity was there. We boarded an old plane that had probably been used by the Khmer Rouge under Pol Pot and his brutal regime. Our taxi, from what was just a field, was by open-top jeep of course so we were covered with red dust, as were our suitcases. We arrived at our hotel, which wasn't a hotel but a disused school. A very large woman carried both our cases on her shoulders up to our room. This seemed strange as there were no other people in the place. The room was bare and there were no amenities at all, not even a kettle to make a cup of tea, and to make it worse, there wasn't a restaurant or any shops and we'd not eaten for ages. We ventured outside to find a place to eat and stumbled onto a small cafe where we managed to get a meal of a kind and got talking to a French guy who lived in the area. He told us of a good hotel a short taxi ride away called The Secrets of Elephants. We stayed one night in the hotel that never was and high-tailed it to see what the secrets of elephants were all about. Well, it was very small with about six rooms, very tasteful and old-world; basic but sweet and homely and just what was required after the hotel from hell.

Next day, we decided to take a kamikaze taxi, a motorbike with two seats at the rear, to the centre of Phnom Penh. After a nightmare of a 'taxi ride', we learned our lesson and hired a car and driver to show us around. He took us to an old school which had been used as a jail and torture centre, where poor sods were tortured and interrogated by the infamous revolutionary Pol Pot, leader of the Khmer Rouge, and his cronies. This was a real hell. The cells were just big enough for one person to sleep on the floor on his/her side; there were photographs on every wall of those unfortunates. We were then shown the torture cells, beds without a mattresses and chains that were used to hold the captives down whilst being tortured and other horrid implements - it made you feel sick looking at and imagining what evil must have taken place. We didn't feel like lunch after that.

After this, the driver took us to the infamous Killing Fields. [I must warn the reader that this next section is disturbing. You may want to skip this paragraph.] The Killing Fields is not a tourist destination but has been turned into a place of gruesome education in memory of the people killed there. There were hundreds of skulls made into statues, and our guide showed us trees with hard serrated leaves used as wood saws to hack off heads of children. As you walked along, it felt like you were walking on pebbles and broken seashells. Our driver informed us that it was in fact human bones that had come up to the surface after the rainy season. We were then taken to a market place. I noticed as we got out of the car that a man had spotted us and was making a beeline towards us. We were tourists and tourists mean money. This man was the most horribly scarred person I had ever come across on my globetrotting and I have seen some sights in my time, recovering lepers included. This was too much and we had to go. Our driver dropped us off at a bar that the press had used during the that awful period. It was one hell of a relief after what we'd seen and a surreal sight to see elephants walking along the pavements by the Mekong river as if it was the most natural thing to do. Back at our hotel, we binned our dusty shoes as it didn't feel right to wear them after walking on human bones.

The Killing Fields, Cambodia.

Viet Cong traps with punji sticks.

We then took a trip down the Mekong river to see the floating villages and crocodile pens. A travel agent told us about a place called Siem Reap, home to one of the largest religious complexes in the world, Angkor Wat, which was not to be missed, so we booked a flight for the next day. As we waited at the airport, which was just an airstrip, this crazy women started to walk towards the taxiing aeroplane which still had its two propellers rotating at a fast rate. I could foresee her being chopped into a number of small portions but fortunately she stopped, as did the propellers, much to our relief and no doubt the pilot's! Our baggage eventually arrived, covered in red dust, and our taxi arrived without a top. So yet again, the luggage wasn't the only things covered in the stuff. We arrived at another dismal hotel in the centre of the town, just a shade better than the one in Phnom Penh but we were only staying one night as we were moving to another place out in the bush. Breakfast the next day was al fresco and we were sat at a table not far from the perimeter fence where a lot of beggars were looking at us with open mouths - not a pleasant thing as you were trying to eat your breakfast, which I found I couldn't finish.

Then it was on to Angkor, which is the site (the Wat bit means temple). This complex was, I'm told, used in an Indiana Jones film and in Tomb Raider with Angelina Jolie. It should be another wonder of the world, it was such an awe-inspiring place. I managed to get a photo taken in-between two monks in their orange robes; they must have thought I was a headcase but it was a wonderful experience. Next, it was on to our latest hotel deep in the bush where the rooms were on stilts. We had dinner outside on the veranda and were served cold beer which went down a treat. The manager informed us that the generator was switched off at night to conserve fuel. The room was rather small, with a double bed and a bathroom and the bed had a mosquito net around it. Sure enough, the generator was turned off - end of lights, so it was a good job

I'd brought a torch with me. As my wife reached into her suitcase, we noticed what looked like a large spider dropping into it. We didn't start rummaging about to find out if we were right; the thing probably accompanied us on the rest of our travels. Needless to say, during the night we were woken by something scratching and making a weird noise on our roof which appeared to be a huge lizard. After my few beers I needed the toilet. Walking to the bathroom was not very pleasant as you didn't know what you could be stepping on and I could see spiders' eyes through the beam of the torch. Spiders don't usually bother me - well, the UK ones don't - but these were as big as your fist and probably poisonous too, so a rather sleepless night was had.

Me between 2 monks, Angkor Wat.

The hotel did excursions, walking trips, so we decided to go on one with a guide to take us through the bush. We soon came upon a small village with huts, also on stilts. Of course we now understood the need for the stilts - what else would have crept into our room had we not been raised off the ground? The people were very friendly and we were invited into one of their huts. The floor was covered in straw and they had an open fire going in one corner of the room. We quickly ascertained where the nearest exit was and kept it in sight as sparks and straw have a tendency to mix very well! A little old woman with no teeth kept offering us food which we politely declined. We managed to grab forty winks as the walk had tired us out, only to be woken by a bloody chicken that was about to peck my big toe. I shot up quick, remembering that bird flu was doing the rounds around the world. This made the old woman laugh heartily: a big chicken afraid of a little chicken. We were later joined by, I would assume, the chief or the village elder, no doubt to see what gifts we had brought him. He was probably ex-Khmer Rouge, judging by his age and his mannerisms. Fortunately, I had a spare pair of soft-napper leather gloves that we wear on the ship when inspecting things, which he seemed to like, going by the smile on his face.

We stepped outside and headed to a stream where villagers were washing and bathing. They had made a chute out of bamboo so the water could cascade down to them like a shower. I joined them and we soaked each other which gave us all a good laugh. Then our guide arrived to take us back to the hotel. Then the fact suddenly dawned on me, and this sent a shudder down my spine, that we were in a place where mines and evil booby traps were possibly still buried and active. I mentioned this to our guide who just smiled and carried on walking, which was very reassuring, and my heart missed several beats every step! I gave him a T-shirt from the ship and some cash for not getting us killed. We stayed one more night in the stilts/sticks and headed to the airstrip, hoping the crazy woman who wanted to be chopped up wasn't there, which she wasn't. At the airport, we got a taxi to a decent four-star hotel where we had a few drinks and looked back on an incredible experience. Then it was back to Kuala Lumpur, where we stayed one more night before my wife flew back to the UK and me to the ship. It had been an eye-opening time, sometimes wonderful, sometimes an ordeal but not your usual holiday. Fully recommended, though, as long as you watch, and respect, where you step.

Water fight with the Hill Village tribes, Angkor Wat.

SYRIA, 2ND TRIP
2009

(Where I say the wrong thing to a Dutchman and star in my very own Stephen King novel.)

My second visit to Syria was on a project for Canadian oil company Oil Canada. I flew into Damascus and was met by a driver from the company, which is always a good sign. So off we set to the main office, where I spent the first week getting to know the personnel and the project, which was in a place in the desert not too far from Palmyra, the ancient Semitic city in present-day Homs Governorate, where archaeological finds date back to the Neolithic period. I had been there several years before when I worked at the University of Mechanical and Electrical Engineering in Damascus, so, as we drove through Damascus, I asked if we could to drive past the place for old times' sake. For lunch, we popped over the road to a restaurant where it was difficult to see because of all the hubbly-bubbly pipes being smoked. It made one feel light-headed just being in the restaurant, so I gave the pipe-smoking itself a miss.

The first afternoon passed quickly. The hotel was on the same block as the office which was very convenient. The first week also passed quickly and I was soon on my way to the site itself, which took three hours to reach. We stopped to stretch our legs by an olive plantation. I was tempted but, alas, didn't have a Martini to go with it. Arriving at the site, I was introduced to my new colleagues who were mainly Canadian and one Dutch guy who I got off to a bad start with as I mentioned the word 'Cloggy'. I know in years gone by, 'Cloggys' was an insult but in the oil industry it's used as a joke, a piss-take and every nationality has their own version. Anyway, he took it as an insult and never forgave me for it. So although we were in the same office he hardly ever spoke to me, which didn't break my heart – I've known many Dutch people and they have always joked about it. I have to take "Scouse git" and "Watch your wallet, there's a Scouse about," so I thought, 'Sod you, Cloggy!' But as his desk was next to mine, it was not a nice situation. Fortunately there was a Jock in there to lighten the atmosphere. The accommodation was an arrangement of Portakabins, which were OK. Breakfast was boiled eggs, pita bread and coffee. Lunch was hummus, salad and pita bread, which I must admit I quite like. There was not much to do in the desert, so after dinner it was back to your cabin to read and sleep. This was a typical construction/commissioning job involving the pre-commission, commission and start up of the plant, which went ahead without many problems. We soon got gas flowing through and the flare was lit up signifying our progress.

One morning, I woke up with a sharp pain in my right side and tried to get dressed but couldn't even tie my shoelaces. I felt nauseous and the dull ache in my tummy quickly became a severe pain in my right side. 'This is appendicitis,' I thought and managed to make my way to the medic's cabin. After knocking for a while, I finally woke him, then tried to communicate that I might have appendicitis and needed to get to a hospital rather quickly. The Syrian medic wasn't quite au fait with my self-diagnosis, with me a) not being a doctor or b) not being Syrian. 'This could be serious,' I thought. 'I could die here with the locality of the place.' I'd been through this before as a boy and had been rushed to hospital then. It wasn't the real thing but the same symptoms, which turned out to be a grumbling appendix, so I had an idea of what it could be. The medic took ages, so I ambled over to the office where Jock and the grumpy Dutch guy were and told the Canadian manager who started to get things moving. Soon, I was in the site ambulance heading for Homs, the nearest town, some 30 miles away.

Lying down in the ambulance was a nightmare. I was not only in pain but very cold as it was early morning and it gets freezing in the desert during the night. It was also a very bumpy ride. Fortunately, a Canadian engineer was with me and managed to hold me down and stop me crashing to the floor of the ambulance. After a rather uncomfortable drive through the desert, we got to Homs, which seemed to be deserted. We arrived at a hospital where nobody could speak much English. Somehow I managed to get through to them with my small amount of Arabic and was taken to a side ward where they preceded to do an ultrasound scan with a portable machine. Next thing I knew I was being transferred to another hospital. Now I was beginning to worry. The pain was getting much worse and I'm thinking: 'If this bloody thing bursts then it's peritonitis and goodbye cruel world.'

Eventually, we got to the other hospital where I was examined by a doctor who could speak English and I heard him tell a nurse to prepare the theatre. I was then wheeled down to the operating theatre. As I lay down on the table, I noticed the anaesthetist bleeding the air out of the hypodermic but the liquid squirted out and landed on my head. She then went on to inject me but I was past caring by then, so long as they got on with it quick. The surgeon was about to proceed when it sunk in: *I was still conscious.* Due to the anaesthetist spilling most of the anaesthetic on my head, I hadn't been given enough and would be able to feel everything. I could still see what was going on and wanted to swallow like you tend to do when having dental treatment but couldn't. I couldn't move, not even my little finger, and could feel them cleaning me with antiseptic prior to cutting. It was going to give me a searing pain, worse than I have experienced in my life no doubt and I wouldn't even be able to scream! This was like something out of a Stephen King book, where you are on the operating table, wide awake and about to be sliced open and you cannot do anything! Luckily for me, the anaesthetist checked my eyes, realised that I was still fully conscious and topped me up with propofol to knock me out.

And so I survived to live another day.

I was in the hospital for three days and received really first-class treatment, possibly due to the fact that oil companies tend to have big bucks but this was none of my concern. I'm just glad that none of this happened in 2021 with this evil war going on due to that bloody murderer Assad. We have just got rid of one evil bastard Saddam Hussein, now we have another, helped as usual by the sneaky Russians. I remember seeing Russians in the desert on my last visit to this place - what was all that about?! When discharged from hospital, I was collected by a driver and driven back 'on the road to Damascus' to our office, which turned out to be another very painful journey. After a couple of days in the hotel, it was back on site. Brute strength had to be used to dismantle a malfunctioning pump and I was about as useful as a chocolate fireguard in hell. So I had to call it a day and fly home. As it was Christmas anyway, my appendicitis had come at the right time but I would not like to go through that scary ordeal again as I'd almost met the Holy Ghost and not many Wise Men, certainly no wise anaesthetists. I wasn't the least bit sorry to come away from this project as I just didn't gel with any of the staff, Canadian or local, so I spent the final week in the main office writing handover reports and just glad to be out of there. So my appendix, which is a useless and non-functional part of the human body, came in handy for me after all!

Me with members of the Islamic State (basically ISIS mark-1).

SINGAPORE, FINAL CONTRACT 2011-12

(In which I get my best job yet and pay the price.)

Singapore. What can one say about this place? It's taken me a while to think of how to start this chapter but here goes... It's my third visit here and also my second contract/project here too. Out of all the countries I've been to in the world my favourite has got to be here. It's such a wonderful place. America is a close second, then Africa. I could go on as each country I've been to has its good and bad things. I've worked in a jungle rainforest, a desert and off shore in the Atlantic Ocean and North Sea. One doesn't get to work in these places every day, does one? Especially a working class lad from Liverpool! Plus it had its dangers too! But for me, Singapore had everything I wanted: it was spotlessly clean - even the dirt is clean - and ultra modern but still retains its culture such as Chinatown and Little India. In the middle of the city, you'd find trees and greenery like a city jungle! There were lots of modern shops, bars and two marvellous quays: Clark Quay, the new one, and Boat Quay, the old one, full of bars, shops and restaurants, great places to go at the weekend. Most places I've been to are great places and some are proverbial hell holes but you can't win them all. I've not mentioned the Middle East as I always felt uneasy there; however, there was always Dubai and Bahrain which were completely different.

For this final trip to Singapore, I tried several apartments before finding a gem on Sentosa, an island off the mainland which you got to by driving over a causeway. It was known as 'Death Island' during the Second World War under Japanese occupation but this was unknown to me during my occupation of my apartment. I'd park the car in my allocated slot in the underground car park, then walk about ten feet to the lift, which came into the hallway of the apartment. On one side of the apartment block was the South China Sea and on the other side was my veranda overlooking a canal with boats going out to sea. This was a magical view when having a sundowner. My job here was as an inspector on a newly built FPSO for the government of Equatorial Guinea, West Africa. The interview was in Liverpool via video conferencing which I'd never experienced before. I entered the room and was facing a huge TV screen. A technician came in and switched it on and we waited for the call from Singapore. I was told I'd be interviewed by two managers of the project, one Canadian and the other Scottish. The call came, but the problem was they could see me but I couldn't see them. It was no problem as it didn't put me off at all and I think I got Brownie points for this. Basically, I was not depending on this contract as I had several irons in the fire and was relaxed and confident. A day later, I received conformation that the position was mine for the taking. So with the salary up to my expectations, the perks top-notch, plus the fabulous location, I had no reservations and accepted their offer. So it was back to Singapore.

The day after my arrival, I took a taxi to the shipyard to meet my new colleagues. At this time I didn't have an apartment, so it was hotels for now which was no hardship and no problem getting to and from work. The taxis were pre-booked so would be there after breakfast and the journey took about forty-five minutes. On my first day, I had a meeting with my Canadian boss; the Scottish manager was at the main shipyard. After the meeting, I was introduced to my new colleagues.

I was to share an office with Kelly who was the structural design engineer, a Chinese-American who lived in Texas with his wife and two sons. Then there was a huge Brazilian guy, Ronaldo, who had his family with him in Singapore. He had to vacate his desk to make way for me. I immediately gelled with them both and they made feel at home, which was a great start to a new job. Mark, the Australian construction superintendent, showed me around the site, which built all the modules with the machinery that would be installed later on the ship/FPSO. I was to inspect these modules prior to their installation and start-up, shake-down and acceptance tests.

The apartment I found was fully furnished, except for cooking utensils and bedding. I had TV stations installed, then got myself a hire car to do my first shopping expedition for groceries etc. I was up and running. Getting to the shipyard meant a drive on the motorway which was pleasant and no hassle with plenty of parking spaces to choose from. After a coffee and a chat with Kelly, I would get my quota of the day's inspections and out into the shipyard I'd go. The personnel at the shipyard were made up of locals, Indians and Chinese. Most spoke English which made one feel a little inadequate. But working abroad a lot, it's surprising how many languages one can pick up, not fluently but enough to get you through, even Mandarin, although Kelly came in handy there with being Chinese! Ronaldo spoke good English too - his mother tongue was Portuguese, coming from Brazil, which he didn't use. At lunch, Kelly, Ronaldo and I would often go to a Chinese or Japanese restaurant or to the yachting-club marina or pick up a sandwich from the local Subway. The day would fly over and soon I'd be dropping Kelly off and back to my place for a shower and a G&T (plural required here). It was really pleasant on the veranda, cupping an ice-cold drink and watching the boats go by.

The weekend started on Friday. I would do my weekly shop, then it was nice coming home to a clean and sparkling place as my cleaner, Mrs Wong, had been in. She would even change the water of my weekly treat of orchids. I would shower and cycle along the South China Sea coast road which was just a cycle path to the local yacht club where there was the ubiquitous Harry's Bar. So it was a cold beer and usually burger and chips whilst perusing the boats, another cold beer then cycle back home. Saturday was another work day but it was only 8am till 1pm. After work, I would go to the local go-kart track and hurl around, sliding and skidding my way for a couple of hours, which was brilliant; drop Kelly off, then home, shower, sundowners and put something on for dinner, then phone a taxi to take me into town, sit on the veranda at Harry's Bar and watch the ladies and ladyboys of the night go into Orchard Towers, which was a complex of bars and nightclubs. You couldn't tell by looking at the hes and shes which was which but it was something different to say the least. Muddy Murphy's was another place to visit as you tend to get an Irish pub in some odd places in the world - there's even one in Baku, Azerbaijan!

Some days, I would have a drink and a chat to an ex-Everton and Liverpool footballer who had been captain of both Merseyside teams and was now working in Singapore as a TV pundit. The job and the life in Singapore was just brilliant and it even staged an F1 Grand Prix, which was the first F1 night-time race. Most Sundays, I'd cook a full English breakfast, then cycle round to one of the beach clubs and have a couple of drinks and some lunch, often cooked oysters, then on to Tim's beach bar – Tim being the owner and a Singaporean – then cycle home again, shower, then cycle to Harry's bar to watch a Grand Prix on TV. Then it was Monday and back to work but you can see how time flew here!

On Sentosa, there were theme parks, such as Disneyland and lots of beaches to spoil yourself on. On occasional Sundays, I'd go to the famous Raffles Hotel and head for the Writers Bar to read the Sunday newspapers and have a few glasses of champagne, or the Long Bar and chuck some peanut shells on the floor. My wife and daughters enjoyed their visits here immensely (my son, Rich, touched down briefly on his way to Australia and was quite content to stay in the bar at the airport!). I'd bought bicycles for them to get around the island on and of course taxis were plentiful and easy to telephone or hail. One novelty, as you were driving along the road in Sentosa towards my apartment, was having to slow down to avoid a peacock as they were plentiful and would fly suddenly in front of the car! I imagine my neighbours were millionaires as they'd be either driving Rolls, Ferraris or Porsches and of course had their boats moored outside their house or apartment.

I have only one bad memory of this time. One night, I'd been to the Hilton for a drink, when I was accosted outside by what looked like two females. 'Ladies of the night', I thought, on the prowl for business. One grabbed my arm and the other grabbed my nether region, ouch!! They were not females but ladyboys judging by their strength and their Adam's apples bobbing up and down. I pushed them away and said, "Please go away" in Scouse. I carried on down the road to hail a taxi and went to checking the time. It was quarter past zero. One moment I had a new Omega Seamaster watch, the next I was asking people the time! The bastards had nicked it from my wrist and I didn't feel a thing - well, in my wrist that is! I obviously couldn't see any sign of them so I went back into the Hilton to ask if they had a recording on one of their CCTV cameras but they didn't. It worked out OK, though, as I was insured and had purchased it at duty-free prices. The quote from a jeweller priced it at eleven hundred sterling more that I paid, so I actually made a profit, but I really was attached to that watch and annoyed at myself for allowing this to happen.

I did land myself in another stupid situation. One weekend, after too many G&Ts, I took a flight to Bali and spent the weekend lounging on the beach and doing some horse riding and nearly broke my neck falling off the horse – no, not because of the DT's with the G&T's. I had ridden before, but I was given a young mustang which was very frisky and which just took off without any warning whatsoever. I managed to hang on trying to look cool but more like a fool and survived to ride again!

The day arrived for the launch of the ship; it had been completed but was in a wet dock. The President, his wife and children were there, as was my wife who was on the conducted tour of the ship and all went according to plan. So that was my last contract/project but what a way to bow out - in a dream of a place, great salary and brilliant colleagues!

It was then off to Russia for a holiday, Moscow and St Petersburg. Moscow had its usual drabness on the drive to our hotel, which wasn't helped by the miserable driver, but the hotel was better than we expected. After breakfast the next day, we were greeted by our tour guide. I hate these guided tours and we preferred to go off on our own but as we didn't know our way around we had no choice. I had been to Moscow before but St Petersburg was amazing with all the museums and art galleries and the palaces were outstanding. I would definitely recommend Russia for its history and culture. I did foolishly sample many varieties of Russian vodka as there are many and I advise anyone to be careful. On our return home we found that our eldest daughter had brought with her from London two gorgeous ginger kittens, he and a she. They had been put outside in the street in a box as the mother was giving birth to them. What type of person would do that?! A friend of my daughter's had taken them in off the street but her own cat objected, so she had to find them a home. We gave them one and they are now eight years old and are brilliant and love living with us.

To end this chapter and book:

I would like to issue a WARNING. The nightmare began the day after we returned from our holiday in Russia. I got out of bed, showered and was about to brush my teeth when I found I couldn't turn the cold water tap on with my left hand. Whilst walking through the house to the kitchen (as I live in a bungalow), I felt very strange and thought, 'I'm having a stroke.' I reached the kitchen and my wife said to me, "What's wrong with your face?" My reply was: "I've always had this face," but that's where the joke ended. She said, "You're having a stroke," and immediately telephoned for an ambulance. Within minutes it arrived and I was rushed to hospital which was, fortunately for me, not too far and had a specialised stroke unit.

After a scan, it was confirmed that I had suffered a major stroke which happened to be one of the more severe types, an hemorrhagic one - 'a bleed'. I was admitted immediately to the stroke unit and my family were told by the consultant that I may not survive the night as only one in two does. Eight years later, I am still here telling the tale so please HEED my warning. Basically, I had overdone it with alcohol and burned the candle at both ends and in the middle too. I remember how and when it may have started: on a weekend trip, I caught a boat to Indonesia and drank too much; I returned to Singapore and the following day headed to one of the beach bars in Sentosa to recover and got into conversation with two Dutch guys. Of course, one bottle of champagne turned into two – I soon found a way to consume quite a bit (a lot) of champagne. At the shipyard the next day, I was feeling a bit (a lot) under the weather and went to see the medic who, after taking my blood pressure, almost fainted as it was sky-high. It was in the severe hypertension area - 180 systolic and 110 diastolic! Another way to look at it, and I am not making any excuses here, one can get too dependent on drink and then being a dick! Acting like a foolish teenager as you are away from home and your family and in another country far away. One can get very homesick and drinking can take the blues temporarily away from you and of course you are drinking with your colleagues and friends until the next day and so it goes on... Believe me, working overseas you can get into this habit very easily and of course no lack of money to go with it too. By all means have a few drinks but think of what happened to me. I have had to retire earlier than I wanted, my left side is paralysed (fortunately I am right-handed), my left leg is dead and feels like wood, and I have severe pins and needles in my left foot and left hand. I am now one-handed, I don't drive any more as I got fed up of crashing into parked vehicles as my vision is impaired on the left side, and I nearly had a third stroke seeing the next insurance quote. The main thing is, I could have killed a child or in fact anybody! And/or me. I therefore have surrendered my driving licence.

So I have become a misanthrope, or in Scouse-speak 'a pain in the arse'. Just ask my family! I am intolerant, ill-tempered, depressed; I can't even tie my own shoelaces, fasten buttons or zips, and ablutions are a bloody nightmare! Having to use public transport is not the best way to get around, plus all the hospital and doctors appointments one is obliged to attend, not to mention the medication one has to take. The big one is: my cognitive functioning is not functioning as it should. I must also point out that my mother and both sisters had strokes and my youngest sister had a brain bleed so it could be in my family's DNA. Only time can tell. So take care and beware, as the Grim Reaper is always watching and waiting.

SORRY ABOUT THAT!!!! "Cheers." Enjoy the remainder of your life and don't let the bastard with the scythe get you too soon. And have one for me or not! But basically, I am not in charge of me any more. Think about it.

One last thing before you help yourself to a drink or two: I have to mention my family and especially my wife, Patricia. Without her help and understanding, I would never have got through this stroke ordeal. If I had been her I would have shot me. So give it some thought, hey?! This said, with all my travels, different contracts and projects experienced and the people that I have met and worked alongside – the good, the bad and the ugly and all the happenings that happened to me – I don't regret it and would do it all again if I come this way in another life!

Patricia and I in New Orleans, 1999.

THE END

CPSIA information can be obtained
at www.ICGtesting.com
Printed in the USA
LVHW072118220922
729069LV00017B/390